AMBIENT VIBRATION MONITORING

AMBIENT VIBRATION MONITORING

Helmut Wenzel
VCE Holding GmbH, Vienna, Austria

Dieter Pichler
VCE Holding GmbH, Vienna, Austria

John Wiley & Sons, Ltd

expensive and budget approaches are limited in certain aspects has to be understood by all concerned parties.

The authors wish everyone applying these methodologies great success and the possibility to share the experience. In order to help to carry out ambient vibration monitoring activities the Green-Eye software is offered free of charge to interested parties. It can be downloaded from the website http://www.brimos.com together with a set of raw data for trial application. The software is designed for data handling and provides the basic features of frequency analysis for ambient vibration monitoring.

Acknowledgements

The advice, support and understanding of fellows, colleagues and family members is required to write a book. The authors are well aware that without this help it would not have been possible to complete it. Many of the examples presented have been taken from actual assessment work performed during the past years. We would like to particularly acknowledge the opportunities given to us by our clients ASAG Innsbruck (Mr. Fink), ÖSAG Linz (Mr. Ritzberger), MA29 Vienna (Mr. Winter), NÖLR (Mr. Talmann), TLR (Mr. Aschaber), KTLR (Mr. Hawranek) and ÖBB (Mr. Glatzl).

Many thanks go to our fellows of the board for their understanding, namely Mr. Peter Fritsch, Mr. Gerd Chiari, Mr. Reinhard Mechtler, Mr. Harald Schmidt, Mr. Christian Eckerstorfer, Mr. Walter Nemeth and Mr. Robert Schedler. Among the many work contributions we would like to highlight the input from external experts of the technical university of Leuven namely Prof. Guido DeRoeck, Mr. Bart Peters and Mrs. Anne Theugels. The work of Yozo Fujino, Dan Frangopol and Emin Aktan also inspired the development described here. Among the many in-house contributors and co-workers we would like to mention Mrs. Bianca Mick and in alphabetic order Mr. Ernst Forstner, Mr. Peter Furtner, Mr. Roman Geier, Mr. Konstantin Savov, Mr. Martin Stöger and Mr. Robert Veit for their work.

This publication would not have been possible without the existence of a number of research projects. In particular the European projects IMAC-Integrated Monitoring and Assessment of Cables (G1RD-CT-2002-09003) and SAMCO-Structural Assessment Monitoring and Control (G1RT-CT-2001-05040) supported works cited here. The Austrian research projects BRIMOS (supported by FFF-Forschung Förderungs Fond der gewerblichen Wirtschaft) and the research project supported by the Ministry of Infrastructure (Dr. Günter Breyer) enabled a development which provided the basis for this publication.

The authors further would like to acknowledge the contributions of numerous owners of bridges, because of their trust in new technologies, who patiently overlooked the initial deficiencies and overoptimistic promises.

Finally we would like to say thank you to all and include all those whom we were not able to mention here.

Summary

Bridges, but also many other structures, have a significant vibration behaviour which may be addressed as 'vibrational signature'. This dynamic behaviour is typical for a structure and can be obtained by appropriate measurements and used for the assessment of the condition of the load-bearing structure and the determination of damages after respective evaluation.

The ambient vibration method (AVM), which is to be presented in this book, is based on the analysis of the dynamic characteristic of structures. Similar to parameter classification from mechanical engineering, where such methods have already matured to standard tools, individual parameters are extracted from the recorded raw data and then interpreted. Measurements from the snapshot up to permanent monitoring produce the required raw data. Due to long-standing development work it was possible to design a hardware system that fulfils the respective technical standards. The theory behind this goes back to the nineteenth century and had several peaks and setbacks. The introduction of computers into practice at the end of the 1980s enabled an appropriate application of this fascinating method.

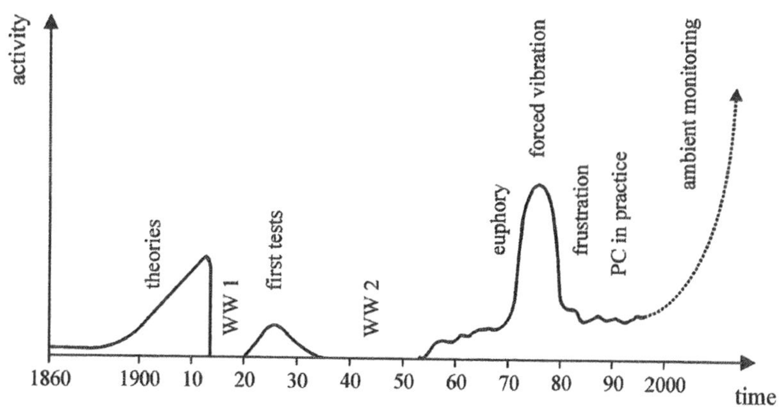

The potential of the method developed is not only limited to the assessment of the actual condition of a structure but also offers a variety of further applications, such as:

- traffic analysis;
- environmental assessment;
- life cycle predictions;
- maintenance planning;
- quality control.

1

Introduction

The increasing number of civil engineering structures in the field of traffic infrastructure entails the increasing significance of maintenance problems of such engineering structures. Bridges are monitored by periodic supervision measures with the aim of minimizing the safety risk for the user, on the one hand, and of keeping the costs for the maintenance as low as possible by the execution of rehabilitation measures at the right time, on the other hand.

Potential Impact

Civil engineers concerned with supervision of structures for safety and maintenance reasons are aware of the limitations of their current practice of condition assessment based on visual inspections. Typical routine applications of condition assessment are carried out on structures applying rating systems. The consequence would be unbearable costs on society for replacement and retrofit tightened up by shrinking budgets. The expressed intention of the bridge owners globally is to reduce the number of bridges rated deficient by 30% within a short time through the application of sophisticated methods based on measurements.

1.1 SCOPE OF APPLICATIONS

Owing to the quickly increasing traffic density, in particular in the field of sophisticated road networks, restrictions of unhindered traffic flow because of inspection works entail high economic costs. Therefore such works must be limited to the absolutely necessary minimum. The ambient vibration method

Ambient Vibration Monitoring H. Wenzel and D. Pichler

(AVM) was developed under the premise that it can be used practically without any impairment of the traffic flow. The aim is to provide a system that makes it possible to reduce the employment of inspection devices to a minimum by well-aimed specification of problem zones, therefore maintaining the traffic flow in as undisturbed a way as possible during inspection works. The procedure can be applied independently of the type and construction of the structure and materials used.

The currently used procedures for bridge monitoring are dominated by manual methods. Visual assessments have a central significance. Devices such as binoculars, magnifying glass for measuring, levelling instruments, endoscopes, etc., are used but the results are dependent on the subjective recognition of damage by the personnel carrying out the examination. It is tried to objectify the recognition by check lists, comparative patterns, etc. Additional methods such as concrete cover measurements and chemically and technical tests help to document an objective image of the maintenance condition.

The AVM has an engineering character comparable with the main manual inspection methods currently used. By measuring the vibration behaviour 'actual' values are obtained, which – with certain restrictions – are not subject to the circumstances of the personnel carrying out the test. The objective condition of the tested structure is determined by a systematic analytical evaluation.

Of course the methods used formerly do not become obsolete but rather provide an additional assessment procedure that improves the application of the method. The examples presented in this report show how AVM can be usefully applied for the support of traditional methods of bridge monitoring. Apart from the requirements for bridge and structure monitoring AVM offers an abundance of further areas of application, which owing to its flexibility are also explained by means of practical examples.

1.2 LAWS AND REGULATIONS

In Austria, according to the relevant legal regulations (federal road law, provincial road administrative laws, building regulations for Vienna), the road administration is responsible for road building and road maintenance. The competent road administration is therefore also responsible for the safety of the engineering buildings in the course of its road network. This results in the following competencies for the Austrian road network:

- federal road network: the provincial governors by way of the indirect federal administration, special companies (ÖSAG, ASAG);
- provincial road network: the provincial governments, the district administration authorities, the mayors.

With regard to road construction and road maintenance no special legal regulations exist. Roads and therefore road bridges have to be built and maintained according to the respective state of the art.

In Austria bridge monitoring in the field of the federal road network takes place on the basis of RVS 13.71, *Straßenerhaltung – Überwachung, Kontrolle und Prüfung – Straßenbrücken* (road maintenance – monitoring, inspection and check – road bridges) at periodic intervals [1]. Here four types of monitoring are distinguished:

- constant monitoring;
- inspection;
- check;
- special test.

Constant monitoring includes the determination of damages that are discernible from outside and is carried out by inspection rides at least every four months. Inspections take place at least every two years with an increased examination depth for the substructure and superstructure by means of a checklist compared with constant monitoring, but usually no special devices are used for it.

The checks (also bridge checks or main checks) are, however, usually carried out every six years. For these checks bridge inspection devices are often used, which frequently require a partial road block. Special tests are carried out if a survey of the actual condition is required due to special events (e.g. after an accident). The scope of such tests corresponds to that of a regular main check; traffic restrictions therefore occur in a similar scope.

In Germany the decisive standard in this field is the DIN 1076, which is valid both for railway and for road bridges. This standard was globally introduced for road construction in all German Federal States in November 1999. For the German railways the D804 with various modules is valid for bridges and other civil engineering structures. They are, however, still being revised at the time of preparation of this publication. In particular, the D80480 is decisive for inspections and/or examinations but must be adapted to the new module system.

In France the decisive standards were published by the Division of Road Management in the Transport Ministry. They are generally known under the title *Technical Instructions for the Supervision and Maintenance of Engineering Structures*. What has to be pointed out is in particular fascicle 31 in part II (1990): *Reinforced and Unreinforced Concrete Bridges*, as well as fascicle 32 (1986): *Prestressed Concrete Bridges*. Fascicle 34 (1986) refers to *Steel Bridges*.

Furthermore, a regulation on the 'Classification of Structures' from 1996 as well as numerous publications and leaflets by LCPC (Laboratoire Central des Ponts et Chaussees), a governmental research institute dealing with building and infrastructure as well as town planning and environmental technology, are decisive. The following titles were analogously translated from French:

- *Handbook for the Identification and Interpretation of Reactions and Signs of Concrete Damages (Fatigue) in Engineering Structures*, LCPC (1998);
- *Examination of Cable Forces by Using Vibration Measurements*, LCPC (1993).

In Italy the decisive standards were published by the Ministry for Public Works (Ministero Lavori Pubblici) in 1971 and 1974. In particular the technical standards for the planning, execution and final approval of road bridges *(Norme tecniche per la progettazione, la esecuzione e il collaudo dei ponti stradali)* need to be mentioned, which were revised and complemented in 1991.

In Great Britain the decisive standards are basically divided into four groups: inspection, maintenance, repair and strengthening, and assessment of roads and bridges. In particular the codes of practice in section one (inspection), including inspection of highway structures and of post-tensioned concrete bridges, as well as inspection and records for road tunnels, and in section four (assessment), assessment of steel, concrete and composite bridges, are interesting.

In this connection the following leaflets as well as a German standard draft are interesting as important aids:

- Automatisierte Dauerüberwachung im Ingenieurbau, Merkblatt des DGZfp-Ausschuss für zerstörungsfreie Prüfung im Bauwesen (AB), dated 12 August 1997;
- Zustandsanalyse mittels modaler Analyse (ZMA) Merkblatt des DGZfp-Ausschuss für zerstörungsfreie Prüfung im Bauwesen (AB), dated March 1998;
- ISO/CD 14963, *Mechanical Vibration and Shock – Guidelines for Dynamic Test and Investigation on Bridges and Viaducts*, from DIN-Normen Ausschuss Bauwesen, dated 12 October 1999.

This standard is to form the regulation on which dynamic bridge monitoring will be based in the future.

Standardization is to be established at a uniform level all over Europe by means of the European network project SAMCO (structural assessment monitoring and control, project number CTG2-2000-33069 of the DG Research of the European Union). This is to be the prerequisite for a quick dissemination of dynamic methods. The progress of the project, which was started in 2001, can be followed under *www.samco.org* on the Internet.

1.3 THEORIES ON THE DEVELOPMENT OF THE AVM

The AVM is a practice-oriented procedure that closes the gap between basic research and application-oriented development by means of statistical methods and approximation methods. Therefore the following assumptions are made:

- The condition of a structure represents itself in the response spectrum.
- The linear static and linear dynamic approaches are not sufficient for describing correctly the dynamic characteristic mathematically.
- The measuring technique has been sufficiently developed.
- Data acquisition and evaluation via computers is efficiently feasible.
- The existing software is appropriate for meeting the requirements.

From the preceding assessment of the status quo the following results can be derived:

- The measuring technique is widely advanced and the data will be lasting.
- The theoretical evaluation can still be developed further.

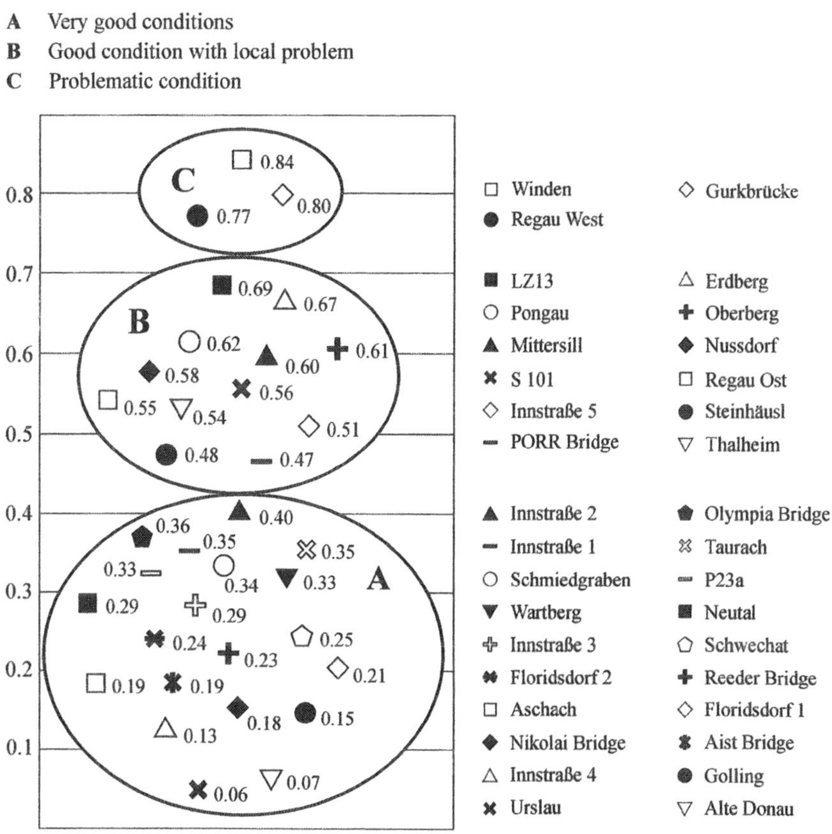

Figure 1.1 Classification of pre-stressed concrete bridges according to the AVM

for checking the calculation assumptions, deviations between measured and calculated results were interpreted and statements for similar future towers were made.

The checking of structures by means of dynamic measuring methods has a long tradition in Switzerland and was carried out until the beginning of the 1990s in the form of tests of free oscillations by means of initial strains or intermittent stresses and by means of excitation with unbalance exciters or hydraulic shakers. Such tests were also carried out in Austria and Germany for scientific purposes, but on a smaller scale, but were not extensively applied for system identification or as a check on for calibration of calculation models. In reference [2], however, further development of dynamic procedures for the assessment of the maintenance condition of structures is suggested.

The rapid development of measuring technology, on the one hand, and computer technology as well as software, on the other, enables us to carry out dynamic measurements of ambient structure vibrations and their evaluation very quickly and with relatively low expenditure. Vibrations influencing the structure due to natural excitation sources like micro-seismic phenomena, wind, waves, etc., are regarded as ambient causes. The measuring and evaluation system BRIMOS® (Bridge Monitoring System) developed by the authors takes advantage of these progressions and opens up a wide field of application to technology.

The dynamic characteristic of the structure is not only used for a single check of calculation models but also for statements on the chronological development of the load-bearing capacity; therefore estimates of the remaining service life duration are enabled by measurements at certain intervals. Measurements taken at any moment supply snapshots of structural integrity and can be used in combination with parallel mathematical analyses for the determination of possible damage to the structure.

The decisive dynamic parameters to be determined for system identification will be described below. During monitoring all analyses of system identification are applied. In addition to the procedures of structure mechanics and dynamics, statistical methods have to be used to determine trends from large data quantities. The use of so-called trend cards (Figure 2.1), which clearly represent an eventual change of individual parameters by means of the time-frequency diagram, has proved successful. This trend card shows the shift of eigenfrequencies over time due to the damage of individual pre-stressed cables in a test at the motorway flyover Regau (Object S123a).

2.1.1 *EIGENFREQUENCIES AND MODE SHAPES*

The eigenfrequencies are an essential parameter for the description of the vibration behaviour of a structure in the linear elastic field. A mode shape – i.e. a vibration

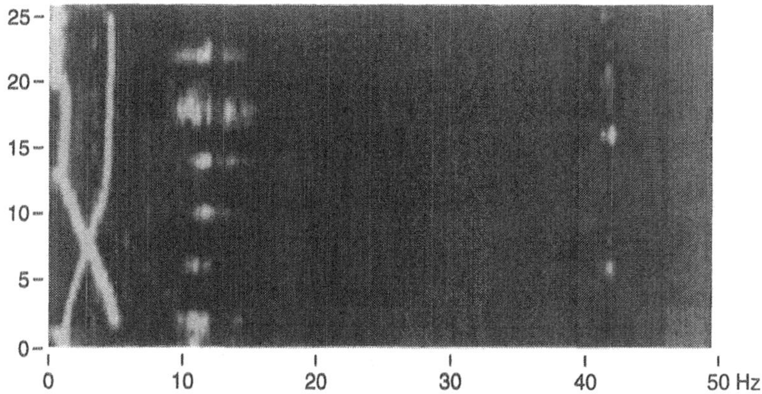

Figure 2.1 Transposition of the system according to damage: moment of failure

form in which the structure oscillates with the respective eigenfrequency – belongs to every eigenfrequency. The actual oscillation of a real structure is composed of the respective shares of the individual mode shapes, such as shown in Figure 2.2.

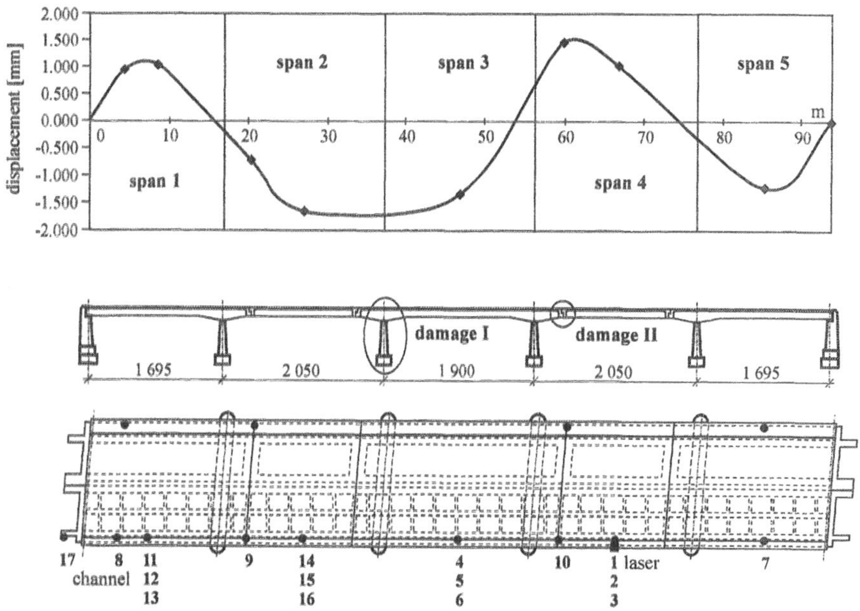

Figure 2.2 Measured first mode shape of a defect structure (settlement of supports due to heavy traffic)

The mathematical modal analysis supplies both the eigenfrequencies and the mode shapes of a structure. In experimental modal analysis the eigenfrequencies shown in Figure 2.3 are obtained as well and the mode shapes can be determined point by point, as illustrated in Figure 2.4 (at the measuring points). Both analyses have to be carried out for system identification. The actual static system is obtained by comparing the measuring results with the calculated values and adaptation of the calculation model to the measurements. In order to get a correct image on the actual load-bearing system, it is required to consider not only the first eigenfrequency and the respective modal form but also higher frequencies and the respective forms [3].

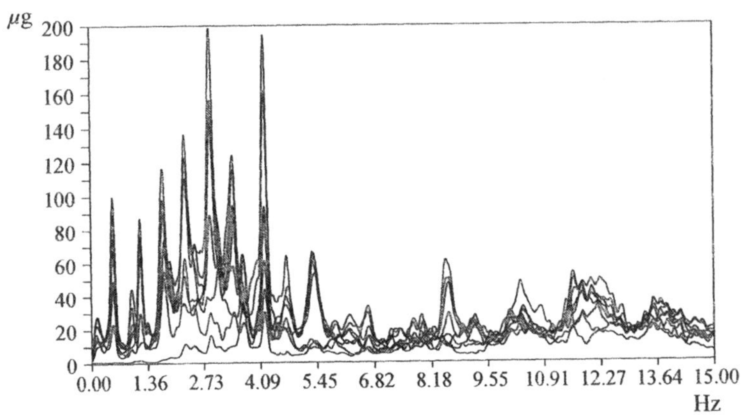

Figure 2.3 Frequency spectrum vertical 0–15 Hz

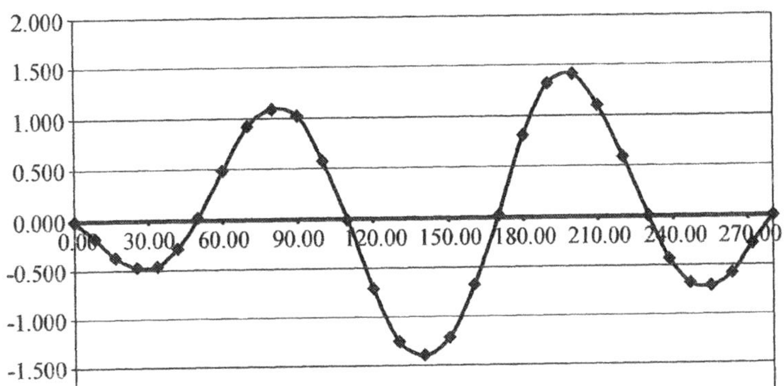

Figure 2.4 First vertical eigenform of a five-span bridge

2.1.2 *DAMPING*

All real structures have a damping that results in a continuous decay of vibrations after excitation until a static equilibrium is reached, as shown in Figure 2.5. The damping properties are dependent on frequencies and represent a significant value for system identification. In particular they are an indicator of the current degree of exploitation of the load-bearing capacity of a structure, as in the case of increasing exploitation of the maximum load-bearing capacity (i.e. at the transition from the elastic into the elastoplastic range, the damping coefficients rise considerably) [1]. In addition, the dampings have an influence on the eigenfrequencies themselves, which is negligible in the case of the usual damping values of structures in use but gains in importance with increasing exploitation of the load-bearing capacity.

In the course of a dynamic test of a structure the determination of the damping properties is therefore necessary to obtain a complete picture on the load-bearing behaviour.

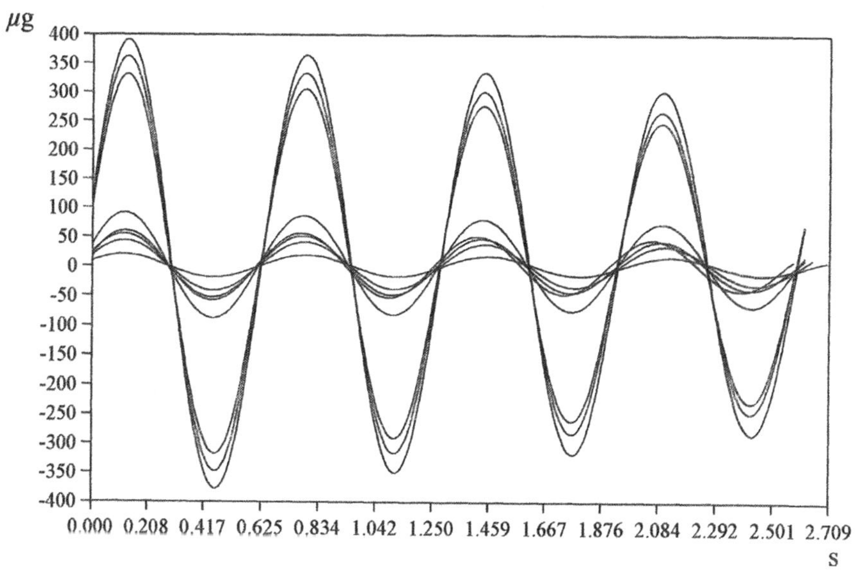

Figure 2.5 Damping window: first vertical eigenfrequency

2.1.3 *DEFORMATIONS AND DISPLACEMENTS*

Up to now the traditional measuring methods mainly measured deformations under defined loads and the latter were afterwards compared to calculated forecast values (e.g. test loads on railway bridges). The information content of this method is essentially limited to the flexural stiffness of the tested structure, but the latter is definitely an important assessment criterion.

Dynamic measurements also include information on the structure deformations during the measuring period. By registering the three-dimensional vibration behaviour it is even possible to extract these deformations from the measurements in all three directions in space. Therefore, not only vertical flexures but also horizontal (transversal and longitudinal) displacements are finally available for the assessment of the structural integrity.

2.1.4 VIBRATION INTENSITY

The vibration intensity is a very good indicator for the stress of a structure by dynamic loads. High intensities of a structure or individual structural members are very susceptible with regard to fatigue-relevant damage mechanisms. The measuring values are recorded and classified in an intensity chart (Figure 2.6). This is appropriate in particular for the localization of detailed inspections at especially sensitive parts of the building. In the case of the Voest Bridge in Linz the decisive sections could be localized on a small part of the total structure. On the other hand, it becomes possible to develop the contribution of the perpetrators (traffic monitoring system) more justly in future by observing them.

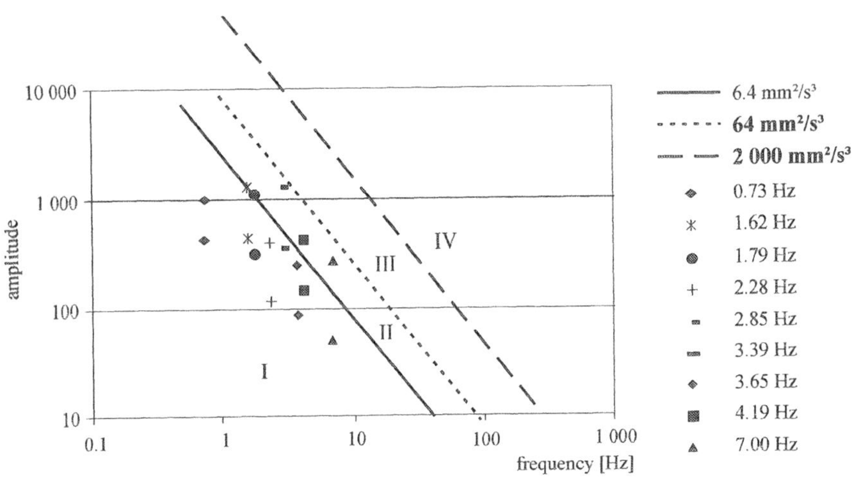

Figure 2.6 Intensity chart at the Europa Bridge of the Brenner Motorway

2.1.5 TREND CARDS

Trend cards represent a signal in the frequency–time domain by means of area mapping. In order to be able to distinguish the individual frequency peaks, colouring of the card is required so that the energy content of the vibration and therefore the respective intensity can be determined. A detailed representation of this technology can be found in section 2.5.2.

As an example from practice such a trend card reveals small changes in a system such as bearing resets. The reset is documented by AVM measurements (Figure 2.7) and laser measurement (Figure 2.8).

2.2 STRESS TEST

Knowledge of the current stress condition of a structure and its individual load-bearing elements is often particularly interesting. Examination is required, on the one hand, to determine the existing current load-bearing safety level and to be able to introduce possibly necessary immediate measures. On the other hand, it is an essential basis for the forecast of future maintenance expenditures. An important assessment criterion to be mentioned in this connection is the evaluation and interpretation of the vibration intensity of the respective structure. By application of classification the endangering of a structure due to damage can be derived.

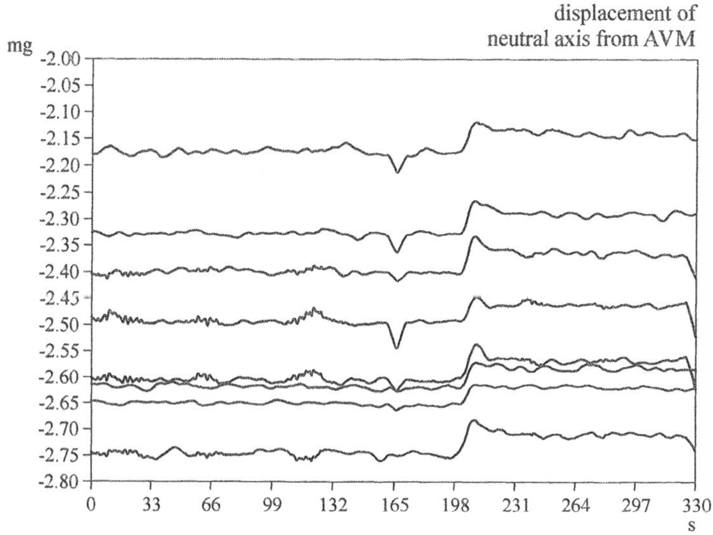

Figure 2.7 System displacement due to bearing reset forces of the flyover at St Marx

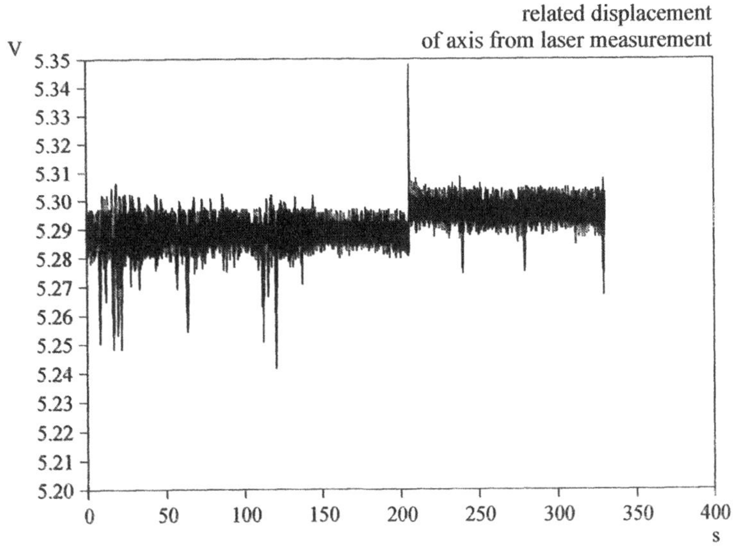

Figure 2.8 Related system displacement due to bearing reset forces of the flyover at St Marx

2.2.1 DETERMINATION OF STATIC AND DYNAMIC STRESSES

The global stress condition has to be determined statically and dynamically. From knowledge of these two stress types the dynamic enlargement ratio for traffic loads can be determined. This is not only an indication of the dynamic sensitivity of the structure; knowledge of it also allows specific measures for the reduction of the stress level (e.g. by speed limits) to be established, as realized on the flyover at St Marx (Figure 2.9).

2.2.2 DETERMINATION OF THE VIBRATION ELEMENTS

Individual elements of a structure frequently show particularly strongly developed vibration behaviour (e.g. rim beams of deck cantilevers, stay cables of guyed constructions). Such structural members have a decisive influence on the convenience for use of a building. They are also particularly susceptible to fatigue fractures caused by continuous changes in stress in the load-bearing elements concerned. Recognition of such structural members by modifying the vibration behaviour within the scope of a dynamic measuring programme enables a timely rehabilitation and therefore prevents greater damage.

Figure 2.9 Flyover at St Marx: section of the A23 motorway

A simple dynamic analysis already supplies information on such elements in a structure and thus enables a specific measurement to be taken. An example is shown in Figure 2.10.

2.2.3 STRESS OF INDIVIDUAL STRUCTURAL MEMBERS

Apart from the global stress situation and the tests on individual structural members particularly susceptible to vibrations, a specific stress determination for individual structural members is required in the scope of an extensive test on the building. All information required for a stress investigation is extracted from an acceleration signal (Figure 2.11). For all load-bearing elements with a direct relation between eigenfrequency and stress level (stay cables, tendons, tension members under compression), determination of stress by means of an evaluation of the vibration measurement is possible. If the relations are more complicated, it is advisable to determine the stresses by means of the optimized calculation model drawn up in the scope of system identification.

2.2.4 DETERMINATION OF FORCES IN TENDONS AND CABLES

Knowledge of the actual tensile forces in the tendons of guyed structures (e.g. pylons, pyramid type roofs) or in the cables of cable-stayed bridges

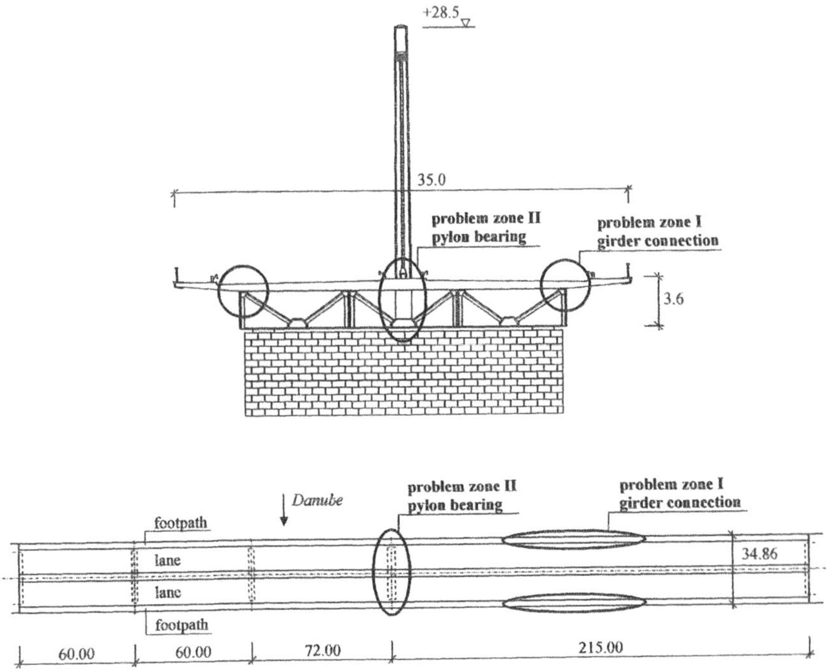

Figure 2.10 Vibration-intensive areas at the Voest Bridge across the Danube [3]

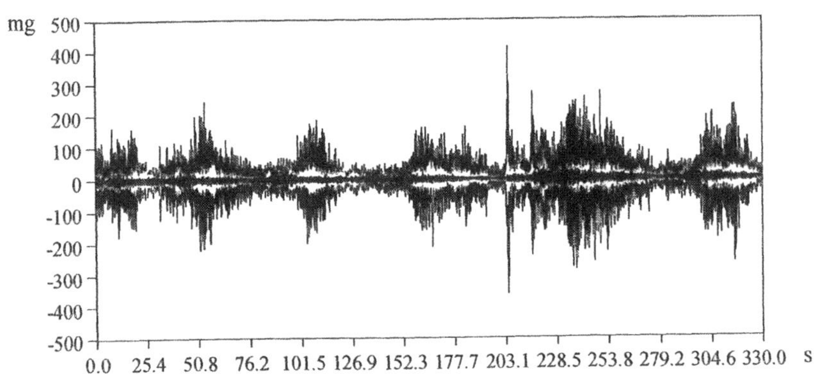

Figure 2.11 Acceleration signal of a bridge cross-section with high values for the cantilever vibration

(Figure 2.12), in the suspenders of arched bridges or in external tendons is required not only for the assessment of these elements themselves but also for examination of the global stress of the structure. Determination of these forces by lift-off tests is connected with considerable expenditure as well as the danger

Figure 2.12 Annual inspection of all cables at the Rosen Bridge in Tulln

of damage. The mounting works at the anchorages can unfavourably influence the durability of these critical elements.

Therefore fast and non-destructive methods for the determination of the forces are required. The measurement of the vibration characteristic is an approach for a solution because there is a simple, quasi linear relation between the eigenfrequencies of a cable and the inherent force.

2.3 ASSESSMENT OF STRESSES

An essential task to be solved is the assessment of stresses of the total system as well as of individual structural members. This assessment must be carried out with regard to the actual structural condition and must consider the predicted future development of the structural condition. The AVM offers the possibility to carry out the assessment on the basis of objective parameters. If these measuring results are combined with calculation models very good predictions can be made by applying probabilistic approaches. Statements with varying degrees of accuracy are possible here and the scope of the executed test determines their expressiveness.

2.3.1 STRUCTURAL SAFETY

The dynamic characteristic contains information on the global structural condition as well as on local phenomena. During assessment of the structural safety the whole system is viewed and its behaviour analysed. The system identification concentrates on the lower frequencies in the spectrum. The scope of tasks is determination of the frequencies and the respective mode shapes of the whole system and calculation of damping values that describe the system behaviour. The total structure can be assessed by means of these results.

For a safety consideration there are numerous excellent models in the literature. Above all the models by Frangopol (USA) and Das (England) [4] should be emphasized. They analyse the service life of a bridge starting from construction up to the slow decline of resistance by use and finally to the so-called intervention points where activities have to be taken to keep the safety level above a critical limit. Such a chart is shown in Figure 2.13. A curve for the theoretical service life can, however, not be transferred to reality. A simple approach is an assessment of the existing load-bearing safety using system identification of the eigenfrequencies. It results in a value – in relation to the planned value – that existed at the moment of measuring. After consideration of the environmental influences as well as the basis of periodic measurements, values can be determined that lead to a curve. Future models will consider further values from parameter classification as well as assessment of the modal parameter eigenfrequency and therefore enable even better predictions to be made.

The aim of these considerations is to find the structures requiring rehabilitation most urgently from a big random sample of structures. The current development of the BRIMOS® recorder, which can register and assess a large number of structures with low expenditure, complies with this aspect.

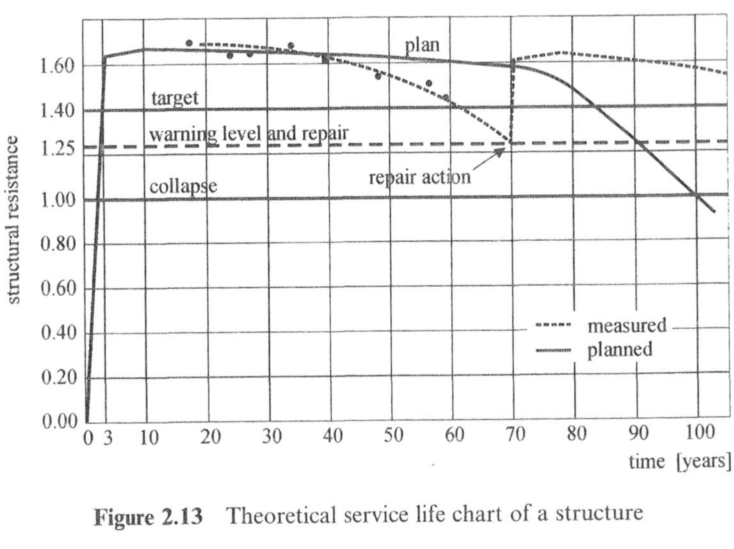

Figure 2.13 Theoretical service life chart of a structure

2.3.2 STRUCTURAL MEMBER SAFETY

The basis of structural safety can also be analogously applied to individual structural members. Every member has its own characteristic which can be isolated from the total measurement. For example, the vibrations of decks are very distinctly discernible. During a sufficiently accurate measurement, i.e. check points at every member, their characteristics can be unequivocally determined. This isolation makes it possible to make statements on the quality of individual structural members. An isolated local damping value is calculated, which gives information on the quality of the respective structural member.

Measurements at several cross-sections of the structure were carried out at the Europa Bridge with the individual elements instrumented according to the sketch in Figure 2.14. Using this method both the load displacement and individual stress can be derived.

2.3.3 MAINTENANCE REQUIREMENTS AND INTERVALS

The increasingly ageing engineering structures in the field of traffic infrastructure entail the growing significance of maintenance problems. Bridges are checked by periodic monitoring measures with the aim to minimize the safety risk on the one hand and to keep the costs for the maintenance as low as possible on the other hand by carrying out rehabilitation measures at the right moment. In Austria the bridge tests in the area of the federal road network are carried out on the basis of the guideline for the monitoring and checking of road bridges at periodic intervals. In order to fix an optimum duration of the intervals, it is advisable to combine conventional bridge tests and dynamic monitoring. On the one hand

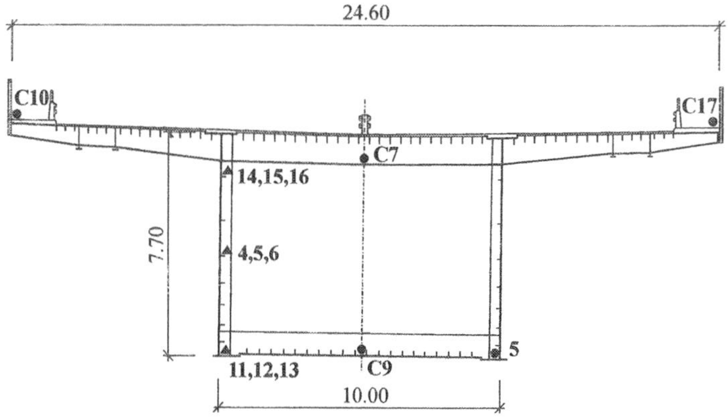

Figure 2.14 Cross-section of the Europa Bridge with sensors

parameters are better and more objectively quantified by dynamic tests; on the other hand also a qualitative statement is possible. This would make it possible to extend the inspection intervals for bridges with a good maintenance condition.

A further attractive possibility is the application of simple verifications, which supply information on changes in the system by means of comparison to the basic measurement. In this way the requirement of a main inspection can be determined at a relatively low cost. So essential maintenance costs could be reduced without having to accept a lower safety level.

Special cases in this connection are the cable-stayed bridges where information (Table 2.1) on the structural condition can be gained by simply monitoring the cables at shorter intervals.

Table 2.1 Results from the frequency measurements for 1997 and 2001 (cable frequencies of Rosen Bridge, shown in Figure 2.12)

	Cables upstream				Cables downstream		
	1997	2001	Difference		1997	2001	Difference
1	2.16	2.16	0.00	1	2.12	2.15	0.03
2	1.77	1.78	0.01	2	1.77	1.81	0.04
3	1.62	1.64	0.02	3	1.63	1.61	−0.02
4	1.38	1.39	0.01	4	1.37	1.37	0.00
5	1.43	1.42	−0.01	5	1.40	1.42	0.02
6	1.27	1.26	−0.01	6	1.25	1.27	0.02
7	1.17	1.16	−0.01	7	1.18	1.19	0.01
8	1.13	1.14	0.01	8	1.12	1.14	0.02
9	1.03	1.03	0.00	9	1.03	1.04	0.01
10	0.99	0.99	0.00	10	0.99	1.00	0.01
11	0.94	0.95	0.01	11	0.94	0.95	0.01
12	0.89	0.91	0.02	12	0.89	0.90	0.01
13	0.87	0.88	0.01	13	0.86	0.88	0.02
14	0.77	0.80	0.03	14	0.76	0.80	0.04
15	0.74	0.79	0.05	15	0.74	0.80	0.06
16	2.21	2.25	0.04	16	2.22	2.27	0.05
17	1.93	1.95	0.02	17	1.92	1.95	0.03
18	1.70	1.71	0.01	18	1.72	1.74	0.02
19	1.48	1.49	0.01	19	1.47	1.49	0.02
20	1.45	1.47	0.02	20	1.43	1.46	0.03
21	1.38	1.37	−0.01	21	1.38	1.38	0.00
22	1.27	1.27	0.00	22	1.27	1.29	0.02
23	1.17	1.19	0.02	23	1.18	1.20	0.02
24	1.16	1.16	0.00	24	1.13	1.17	0.04
25	1.11	1.12	0.01	25	1.09	1.12	0.03
26	1.02	1.03	0.01	26	1.02	1.04	0.02
27	1.04	1.05	0.01	27	1.03	1.04	0.01
28	1.01	1.02	0.01	28	1.03	1.03	0.00
29	0.98	0.99	0.01	29	0.98	0.99	0.01
30	0.95	0.97	0.02	30	0.96	0.98	0.02

2.3.4 REMAINING OPERATIONAL LIFETIME

For the maintainer and the user of a structure the question of the expected remaining operational lifetime is of essential significance. As already described in section 2.3.1, the remaining operational lifetime can be extracted from measurements at periodic intervals. Safe forecasts are possible on the triple period of the measurements, i.e. in the case of measurements over three years, an extrapolation on nine years. From this point of view it seems important to carry out first measurements as soon as possible, even if there is no immediate reason (problem) for them. This type of application of the system will be significant only after several years of employment.

2.4 LOAD OBSERVATION (DETERMINATION OF EXTERNAL INFLUENCES)

The objective of the determination of external influences (also called load observation) is the complete registering of traffic loads or other influences acting on the structure. In this connection the induced loads are not registered by means of a special balance but by the dynamic reaction (response) of the structure. This requires knowledge of the dynamic system behaviour of the structure as acquired by model calculations and/or experimentally (measurement). The dynamic reactions and the determined traffic loads can be used for the treatment of further tasks.

2.4.1 LOAD COLLECTIVE

This is the determination of load collectives by recording the acting stresses according to location, type, size, duration and frequency (Figure 2.15). The influence of wind and temperature can also be considered. Sophisticated load models allow more precise statements on the service limit state and remaining lifetime of damaged structural members.

2.4.2 STRESS CHARACTERISTIC

Load models that can only be estimated with difficultly in the design phase (above all special structures like cable guyed structures, suspenders or lateral braces of bridges) can be essentially improved due to measuring data. Knowledge of the stress characteristic enables improved and clearly more efficient designs.

An example is a railway bridge for which the load displacement was followed by measurements, as shown in Figure 2.16. The reason for the displacement

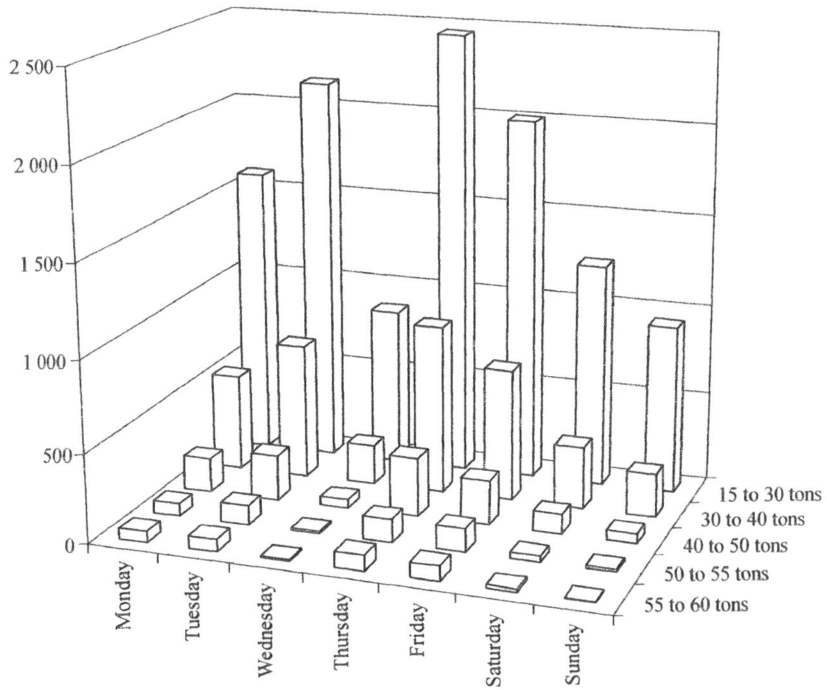

Figure 2.15 Passage events during a week (4 to 10 January 1999)

Figure 2.16 Detailed examination at the Rohrbach Bridge

was the particularly high noise radiation of the structure because no ballast existed. The contribution of the individual elements to the total sound spectrum was to be determined. The aim was to apply damping measures at the right location; therefore the vibration-intensive structural members were to be determined.

2.4.3 VERIFICATION OF LOAD MODELS

A further application of the method is to check how far the theoretically applied load models correspond to reality. This is especially important for older structures where the standards required evidence for lower traffic loads. The proof of whether the structure is up to the current actual loads can be done on the basis of dynamic measurements. Furthermore, the method can be applied for an assessment of widening of the structure. The permanent observation of the traffic data by measurements, as, for example, traffic intensity and density as well as conclusions made on the total vehicle weights, become increasingly significant for strategic planning. A further possibility is the monitoring of important roads with regard to overloaded vehicles.

The actual dynamic factor affecting the structure is dependent on several parameters, but essentially, however, on the driving speed, the vehicle weight and the pavement condition. From the measurement the respective values can be extracted and included into a security check (Figure 2.17). One of the most

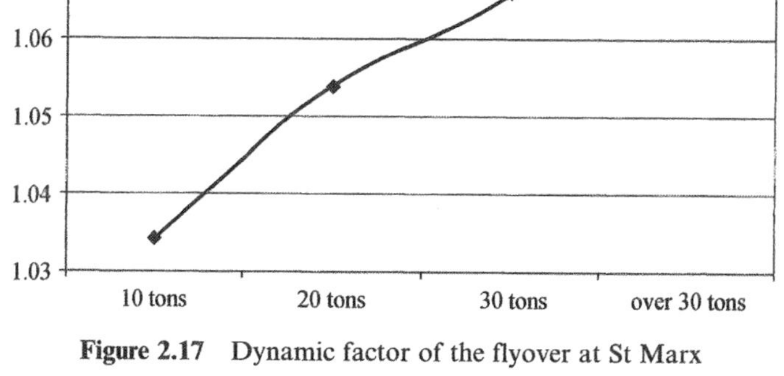

Figure 2.17 Dynamic factor of the flyover at St Marx

important findings of the tests was the fact that in most cases the actual dynamic factor is considerably lower than prescribed by the standard. For individual structural members, however, considerably higher factors were discovered (e.g. 1.90 for the cantilevers of the Europa Bridge). In particular for old structures where the standards still prescribed relatively low stresses, such an assessment is often the last possibility for being able to prove the load-bearing safety.

2.4.4 DETERMINATION OF ENVIRONMENTAL INFLUENCES

Conclusions on the influences from the environment, like aerodynamic vibrations, running water, shocks from neighbouring industrial plants or other excitations, are enabled by dynamic measurements. An external influence often leads to unusual behaviour of a structure. In areas prone to earthquakes it is possible to determine dominant site frequencies, which are to be avoided in the design of the structure.

The influences from the environment are determined by means of an external sensor in the scope of the measurement. This sensor is located outside the system and measures the vibrations transferred via the ground. This has proved particularly successful in the vicinity of a quarry (Mur Bridge Kraubath). Registering the ground spectrum is regarded as one of the most essential development steps of the system considering above all the safety of buildings with regard to the introduction of Eurocode 8. In the scope of a big project in West Java (Indonesia) the viability of this approach for an earthquake mapping based on dynamic measurements has been proven.

2.4.5 DETERMINATION OF SPECIFIC MEASURES

In the case of problems or changes of the requirement (e.g. extensions) the method can be applied for the deduction of specific measures for stress reduction. This can be realized by reducing the stress or changing the construction, thus enabling utilization of still existing reserves. An example to be mentioned in this connection is the structure of the F9 Donnergraben Bridge on the Tauern Motorway, where a damaged concrete deck led to increased vibration intensity and high damping values (Figure 2.18) at the structure. If this value is outside the safe area, measures need to be taken, e.g. the rehabilitation of such defective spots. The scope of the measure can be localized very specifically by examination.

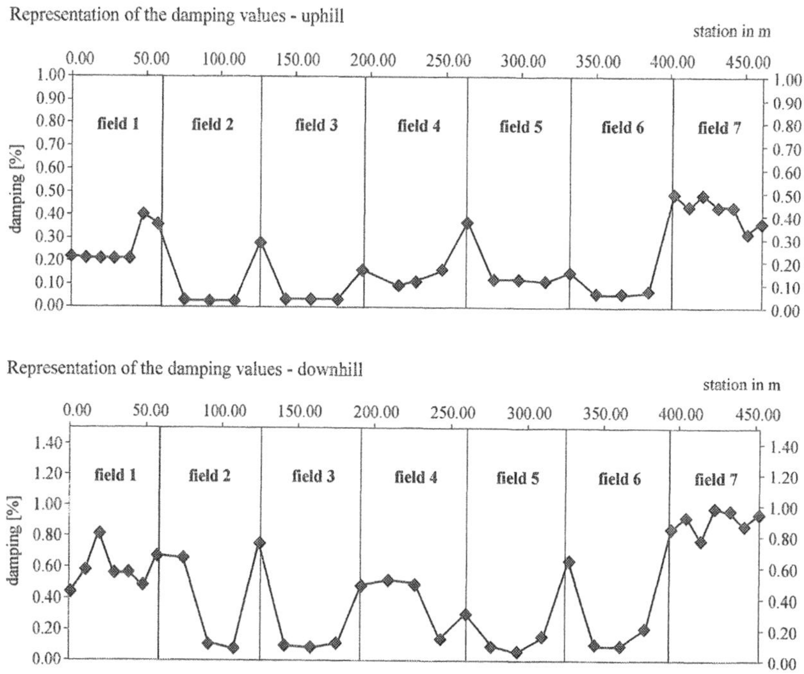

Figure 2.18 Damping progress of the F9 Donnergraben Bridge

2.4.6 CHECK ON THE SUCCESS OF REHABILITATION MEASURES

By taking measurements before and after rehabilitation information on the quality of the works and the success of the rehabilitation can be gained. This is possible by a simple comparison of the dynamic characteristic. An example to be mentioned in this connection is the Inn Bridge Hall-West, where a clear deviation from the system was noticed during the basic measurement. This could be attributed to the unintentional fastening of the system in the area of the abutment. After the removal of this fastening the calculated and expected frequency value was measured in the verification measurement. The measurement could therefore be classified as successful and was correspondingly proved (Figure 2.19).

2.4.7 DYNAMIC EFFECTS ON CABLES AND TENDONS

The dynamic behaviour of a cable is clearly defined by its free vibration length, cable mass and the inherent force according to the cable theory if the cable is

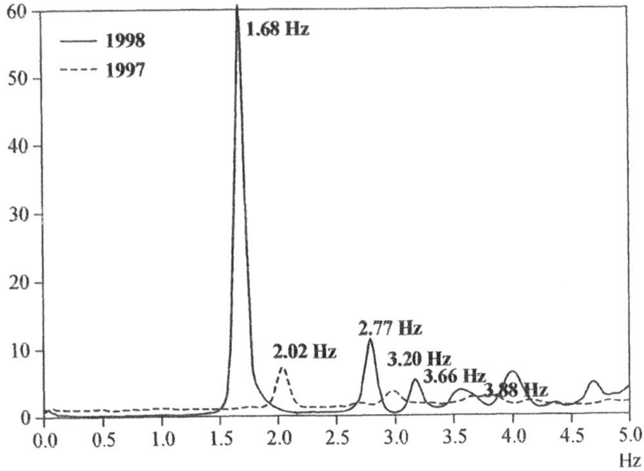

Figure 2.19 Frequency spectrum of Inn Bridge Hall West (1997 and 1998)

without any bending stiffness. Since a typical cable used in civil engineering has a considerable stiffness a method was developed for AVM cable force determination that was partly based on the beam theory. Bending stiffness increases the frequencies, especially in higher modes, because higher curvature in the mode shapes causes a nonlinear relation between eigenfrequency and its respective order (Figure 2.20). Furthermore, the stiffness of a cable shortens its free vibration length (effective length) owing to fixed support conditions entailing an increase of the eigenfrequency in all orders.

Since bending stiffness is unknown for almost every measured cable but consideration of bending stiffness is indispensable for exact cable force determination, the stiffness has to be determined in analyses of the dynamic characteristics before actual calculation of the cable force can take place. The method used for the AVM cable force determination was developed within the European project IMAC (integrated monitoring and assessment of cables), where not only a high number of cables were measured in field tests but also investigations and verification on tendons and stay cables were performed under well-known laboratory conditions.

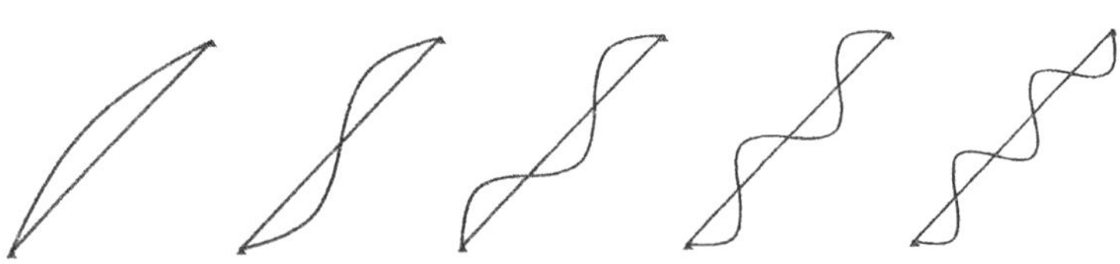

Figure 2.20 Increasing curvature in rising mode orders

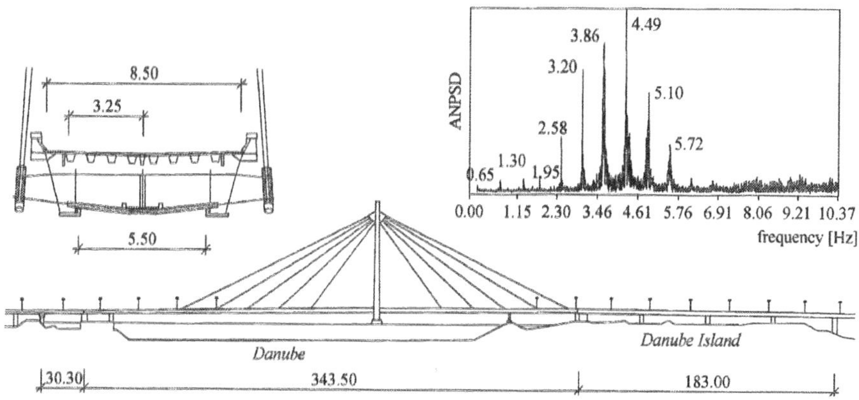

Figure 2.21 Characteristics of cables of the Donaustadt Bridge (all dimensions in m)

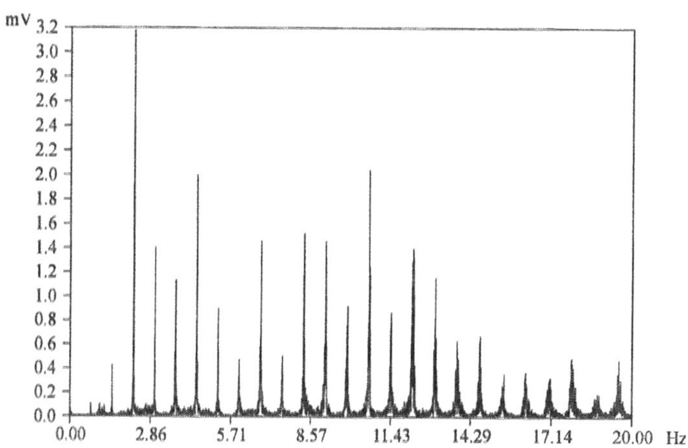

Figure 2.22 Spectrum of a cable of the Donaustadt Bridge

Determination of cable forces enables accurate, quick and cheap quality control under construction, after pre-stressing procedures as well as periodical supervision for safety and maintenance reasons have been carried out (Figures 2.21 and 2.22).

2.4.8 PARAMETRIC EXCITATION

Problems with the vibration behaviour of individual structural members, like cables, struts or delicately proportioned elements, can often be attributed to parametric excitation. In this process the eigenfrequency of a structural member is stimulated by frequency of the total structure. Consequently, an uncontrolled vibration of cables or a fracture of individual stays in steel construction occurs. These phenomena can be quantified by measurements so that the assessment basis for rehabilitation can be derived.

Parametric excitation was particularly observed in the higher eigenfrequencies of cables. Therefore it is not sufficient to harmonize the first eigenfrequencies (basic frequencies) of the cables with the deck frequencies. The fact that cables permanently oscillate (as, for example, in the case of the Erasmus Bridge in Rotterdam) is attributed to the constant supply of energy via parametric excitation at a higher eigenfrequency. The cable vibration itself mostly takes place in the second eigenfrequency with superimposed, higher frequencies. It is, however, inadmissible to draw conclusions on the energy transfer from the visually recognizable vibration form.

The possibilities and examinations mentioned up to now do not represent the whole potential of the dynamic vibration method for cables and tendons. In future further phenomena will probably be identified and discussed. An extension, above all on the behaviour of road joints, has already occurred and other members, which are frequently damaged, are included. Therefore a continuously growing extension of the application field is expected.

2.5 MONITORING OF THE CONDITION OF STRUCTURES

During health monitoring of structures global and local structural properties are assessed on the basis of continuously recorded measured variables. It is therefore possible to predict further development of the structural condition sufficiently accurately. An additional aim is to provide simple and quick identification and recording of changes in the load-bearing behaviour.

The strategy for the development of the AVM considers the following parameters:

- The observation covers long periods.
- Many objects are to be monitored.
- The overwhelming number of objects are in a very good or a good condition.
- A distinction must be made between scientific and commercial applications.
- The system must be very flexible to enable various tasks to be processed.

From these stipulations it becomes clear that the system must be mobile and easy to operable. Permanent monitoring is only applied for special cases or certain problems.

2.5.1 ASSESSMENT OF INDIVIDUAL OBJECTS

In order to assess the structures objectively there is a need to introduce examination categories that differ in the necessary examination expenditure and therefore in the subsequent costs. For the assessment basically four categories are available, which refer to a beam structure with a length of approximately 300 m. For structure types such as cable-stayed bridges, arched bridges or special structures, as, for example, dams or superstructures, a respectively higher expenditure is expected. Therefore a certain bandwidth for the required time period is given for the examinations (Table 2.2).

According to the fixed time expenditure, which is oriented according to the desired information content of the examination as well as to the size of the structure, varying costs for the four categories are to be stated. What should be mentioned, however, is the fact that only approximate values can be stated. This is because the costs of the different systems as well as the marginal structural conditions refer to the standard structure of 300 m (Table 2.3). A detailed cost statement is carried out in the course of the tender preparation and depends on the structure type.

The accuracy of the examination is to be seen with regard to the sensor density for system identification as well as the number of instrumented single components. The examination of category I can be regarded as a standard examination where the individual assessments given in Table 2.4 are carried

Table 2.2 Definition of the examination period

Category	Examination period
BRIMOS® recorder	Hours
Category 0	1 day
Category I	1–2 days
Category II	3–5 days
Category III	Weeks–months

Table 2.3 Costs for the different examination depths (basis 2000)

Category	Costs (euro)
BRIMOS® recorder	≈2 200
Category 0	≈8 500
Category I	≈13 000
Category II	≈55 000
Category III	≈145 000

Table 2.4 Survey on the individual examination depth

Examination category	Individual analyses during the examination
Category 0	General analysis of the structure (visual inspection, sifting of documents, preparation of test plans) Ambient acceleration measurement of the structure Quality control of the records and short *in situ* evaluation, plausibility check of the data Detailed evaluation of the data (eigenfrequencies and mode shapes from measurements, determination of system damping, material damping and vibration intensity) Preparation of a short report including recommendations and measure catalogues
Category I	Preparation of a finite element calculation model Detailed evaluation of the data (eigenfrequencies from the measurements, comparison of the mode shapes model: measurement, comparison eigenfrequencies model: measurement, system damping, material damping, assessment of vibration intensity), MAC (modal assurance criterion, see section 5.8) factor System identification and calibration of the calculation model Preparation of the report including findings, required measures as well as recommendations Damage scenario if required Ambient acceleration measurement of the stay cables (optional) Ambient acceleration measurement of external tendons (optional)
Category II	Instrumentation of piers and foundations Periodic verifications of the structure Comparison of basic measurement with subsequent measurement Instrumentation of local areas like individual cross-sections, piers, special components like links and construction joints
Category III	Examination of local areas like cross girders and single structural components Installation of a fixed measuring system Establishment of a connection for remote control as well as maintenance of the total system Long-term assessment of the structure Video monitoring of the traffic events at the structure Selection criteria for possible trigger control Establishment of an automatic warning system for critical structural conditions

out. It has to be noted that the examinations stated in a lower category are contained in the next higher category.

2.5.2 PERIODIC MONITORING

The most efficient employment of the system is the execution of measurements at variable periods depending on the condition of the structure. After the basic measurement where the system is unequivocally identified, only verifications are required to test the system for changes. The requirements for further tests can be concluded from the results. The measurement of 200 m of structure per day has resulted as the average value for the more detailed basic measurement. In the case of verifications a 500 m long structure can be covered in one day. Figure 2.23 shows a measurement taken on a railway bridge.

2.5.3 BRIMOS® RECORDER

The development of so-called trend cards (Figure 2.24) can be regarded as an essential result of the damage tests as well as of demolitions of these artificially damaged structures in 2001. These cards represent the signal in a frequency–time range by choosing a two-dimensional surface representation for reasons of

Figure 2.23 Verification at the ÖBB Bridge Tulln

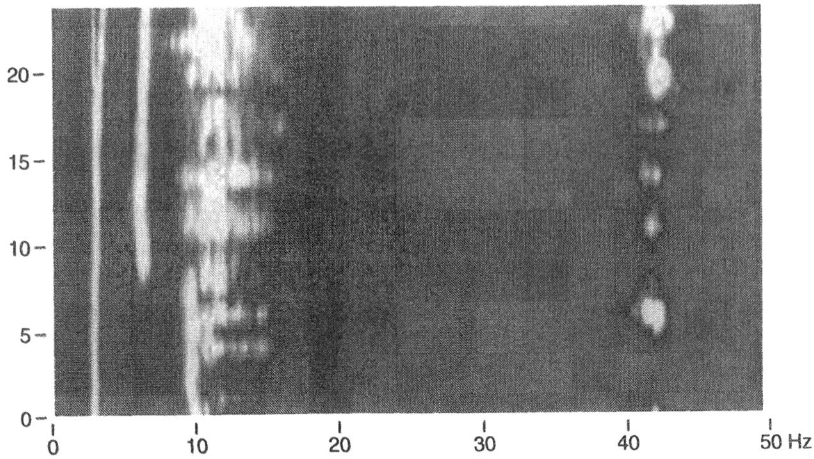

Figure 2.24 Trend card

clearness. Figure 2.25 shows the preparation of the respective trend cards by evaluating the results (frequency spectra) of several measurements, telescoping them together and viewing them from above (Figure 2.24).

In order to be able to distinguish the individual frequency peaks, colouring of the card is required so that the energy content of the oscillation and therefore the respective intensity can be determined. This type of representation shows that damage is already visible in the beginning phase of the frequency spectra.

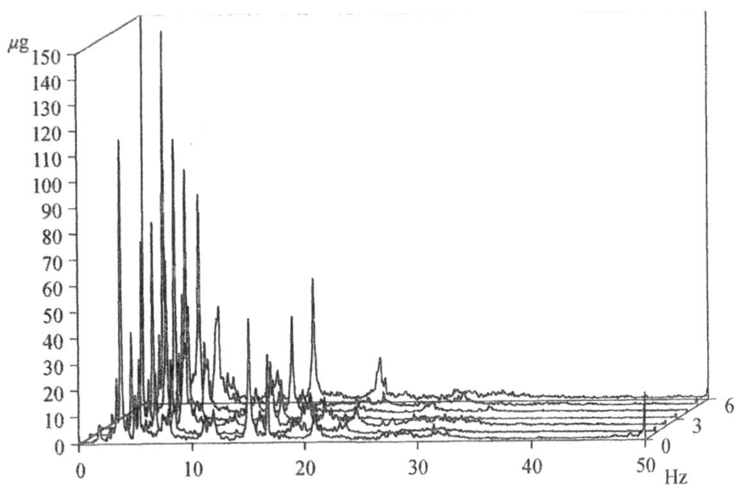

Figure 2.25 Generation of a trend card

What has to be mentioned, however, is that the basic frequencies with their long-wave vibration forms are insensitive to local damage. Therefore the assessment and interpretation of the whole measured frequency spectrum assumes a greater significance.

A generalization with regard to the assessment of the respective frequency band is, however, not admissible, as the behaviour is strongly dependent on the structure type. Consequently, local damage, e.g. on larger structures, could bring about clear changes in eigenfrequencies and the respective basic vibration forms.

The accompanying dynamic examinations at several structures demolished in 2001 [5] clearly showed that one sensor, mounted at a favourable spot on the bridge, already supplies a very high information content with regard to possible damage (long-term observation). A point on the greatest span of the structure has proved a very favourable position; here changes in the dynamic characteristic of the structure become clearly visible. For application of the BRIMOS® recorder therefore the following rule can be applied for installation (Figure 2.26):

$$S = 0.4L_{max} \tag{2.1}$$

where

S = station of the BRIMOS® recorder measured from a column of the main span

L_{max} = maximum span of the structure

For development of a procedure for periodic supervision (long-term control) of the system therefore a system has to be established that works on the basis of the following criteria:

- A sensor at the right location registers damage early and reliably.
- An easily operable and robust unit is required (BRIMOS® recorder).
- The price of the unit should be reasonable.
- The files are read by a laptop or PC and sent for assessment and interpretation by e-mail.
- The interpretation and representation is carried out by an expert system.

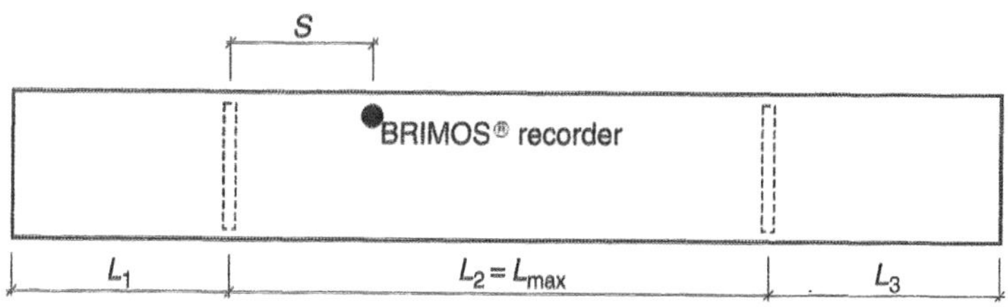

Figure 2.26 Installation of the BRIMOS® recorder at the structure

In order to assess the chronological development of the structural condition dynamic measurements of the structure are required to be carried out at periodic intervals. In this context a six-monthly examination interval seems useful, but at the beginning of the measuring series a sufficient number of basic values have to be collected over a shortened interval. Based on these basic values the chronological development of the condition can be represented graphically by trend cards. The following findings have to be stated:

- Proof of damage of cables is possible long before visual signs, in particular by the trend cards, become evident (minor damages).
- The method works for all structure types (non-pre-stressed, pre-stressed). This was proved by reference tests.
- The BRIMOS® recorder represents an economical possibility of structure supervision and classification.

2.5.4 PERMANENT MONITORING

For special cases permanent monitoring should be taken into consideration, namely if there is good reason to doubt the load-bearing capacity of a structure and where minor changes can lead to serious consequences. In this connection it is important that alarm data are immediately reported to the headquarters via telemetry.

A further important aspect of permanent monitoring is the preparation of statistics on actually existing traffic and its effects on the structures. On this occasion additional video monitoring (Figure 2.27) has proved successful in order to obtain an optical impression of the load triggers. Such a system was installed on the Südosttangente A23 at the St Marx junction. Temporarily, approximately 200 triggers (much too fast and/or overloaded vehicles) per day were registered.

Figure 2.27 Fixed system at St Marx and video supervision

2.5.5 SUBSEQUENT MEASURES

Statements which are assessed and result in specific measures arise from the results of monitoring. During health monitoring of global structural properties the current structural condition is assessed by applying the following methods:

- strain monitoring;
- displacement monitoring;
- curvature and tilt monitoring;
- observation of selected frequencies of resonance;
- selective monitoring of decisive vibration forms.

This enables conclusions on settlements of supports, global stiffness changes, changes of boundary conditions, system changes and the function of individual parts of the structure.

Local structural properties are checked if a local damage exists or a structural member is exposed to special stress conditions. If in the scope of monitoring the defined reference values are exceeded, further detailed tests are required. Examples of monitoring local structural parameters are, among others:

- progress of length and width of known single cracks;
- observation of structure areas susceptible to cracks;
- observation of known weak points;
- extension at points with a high stress concentration;
- flexures and vibration amplitudes of structural elements;
- settlement of supports;
- scour development;
- tendons external;
- behaviour of bearings and road joints.

As the structural change of individual members often only has local representation, the success of health monitoring is dependent on the correct arrangement of the measuring points.

2.6 APPLICATION OF AMBIENT VIBRATION TESTING TO STRUCTURES FOR RAILWAYS

Railways are one of the most important means of transportation for the future. One reason for this is the very effective use of energy with low damage to the environment. Possible very heavy loads, high speeds and an automatic track without the use of human drivers, and therefore economic advantages, are further reasons. These advantages will lead to a renaissance in railway transportation systems over the whole of Europe.

Structures for railways are characterized by high dynamic loadings. Therefore the dynamic response of such structures is of major importance. This section focuses on special structures relevant to railways as a permanent way and related structures.

Owing to the high impact of railway loads, at least for all axle-sensitive elements of the permanent way, a dynamic design procedure is needed. The input values for these procedures need to be measured. Therefore, for a long time dynamic measurements have been performed. The elements of railway superstructures are the rails, the sleepers, the ballast bed or – in the case of a ballast-less permanent way – the track slab and the subsoil.

Depending on the distance of the element to the wheels of the bogies, the dynamic response shows different characteristics. For rails and sleepers the frequencies of excitation are in a broad band between 0 and about 2000 Hz; the acceleration level is very high and reaches values up to 30 g for concrete sleepers. Due to the damping characteristic and the natural frequencies of the elements of the superstructure the transmission of the vibrations into deeper parts of the superstructure leads to a change in frequencies and the acceleration level [6].

2.6.1 SLEEPERS

The design of sleepers is based on European Standard EN 13230 and on UIC-Codex 713 E and takes the dynamic impact into account. For regular concrete sleepers this dynamic impact is well known and in the quasi-static design procedure these effects are considered by using two kinds of enlargement factor: regular dynamic increment factors γ_p, γ_v and γ_i depending on the damping of the rail fastening system, on the speed of the vehicles that are expected to be running on the track and on uncertainties in the contact between sleeper and ballast bed. The second group of enlargement factors k_1 and k_2 cover extraordinary or accidental dynamic impacts. The total bending moments for concrete sleepers are given by the following formulas for three levels:

- level 1, basic design moment: M_{dr};
- level 2, extraordinary moment: $k_1 \times M_{dr}$;
- level 3, ultimate moment: $k_2 \times M_{dr}$.

The starting point for the calculation of the basic design moment M_{dr} is the design value of the load on the sleeper:

$$P_d = \frac{Q_0}{2}(1 + \gamma_p \gamma_v)\gamma_d \gamma_r \gamma_i \tag{2.2}$$

where
Q_0 = static load
γ_p = dynamic increment for damping effects
γ_v = regular dynamic increment (covers speed effects)
γ_d = load distribution factor
γ_r = partial safety factor (covers uncertainties in the contact sleeper – ballast bed)
γ_i = dynamic increment for uncertainties in the contact sleeper – ballast bed

The values γ_p, γ_v and γ_i are well known for regular concrete sleepers and are given in EN 13230-1 and UIC-Codex 713 E. For the development of new kinds of sleeper dynamic measurements of the dynamic impact are essential.

First approximations of the dynamic impact had to be done without measurement results. After realization of the first test track the dynamic impact resulted in unexpected cracks in the frame corners of the sleepers. Measurements of the acceleration level showed that the dynamic impact is essentially higher than expected. Instead of a maximum acceleration level of about 30 g for conventional mono-block concrete sleepers the accelerations increased on frame sleepers up to 50 g. These measurement results led to a new design of sleepers to significantly reduce the high acceleration level. Further test tracks with this new design show that now no further problems with cracks occur and the new dynamic design is successful.

Today booted concrete sleepers are becoming more and more important. An elastic coating or an elastic layer on the bottom surface of a concrete sleeper leads to an increase in the effective contact area between the sleeper and the ballast and therefore to a decrease in the maximum stresses in the ballast bed. Furthermore, the coupling between the sleeper and the ballast is changed by elastic booting.

The elastic characteristic of the coating or layer may cause the dynamic characteristic of the sleeper to change. Whether the dynamic impact load increases or decreases depends not only on the characteristic of the elastic element on the bottom of the sleeper but also on the function of the elastic element between the rail and the sleeper.

Using dynamic measurements on a test track in Upper Austria near Timelkam (Figure 2.28) provides an impressive example of what can be discovered using ambient vibration testing. The test track consists of several different kinds of sleeper. Measurements are taken of the dynamic response of the different systems; conclusions for the design are possible by comparing the acceleration levels. The different sleepers are named K1 for conventional prestressed mono-block concrete sleepers and RS95 for the new frame sleepers. All elastic coatings are products of Getzner, based on polyurethane. The abbreviation SLS stands for Sylomer® and SLD stands for Sylodyn®. The number gives information about the static rigidity and thickness of the coating; for

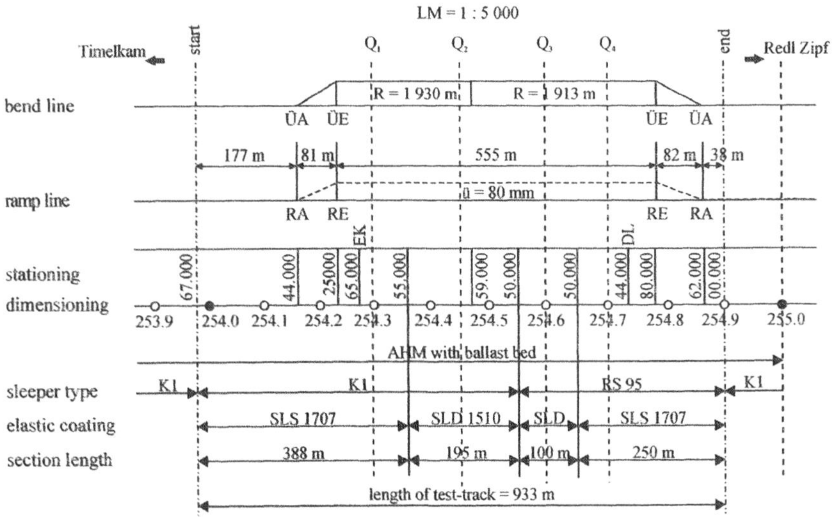

Figure 2.28 Layout test track at Timelkam

example, the number 1707 means a rigidity of $0.17\,\text{N/mm}^3$ and a nominal thickness of 7 mm [7].

The natural source of vibrations was the regular run of trains on the test track, which is part of the Austrian Westbahn line. The results for the acceleration level for the different kinds of sleeper are given in Figure 2.29. The

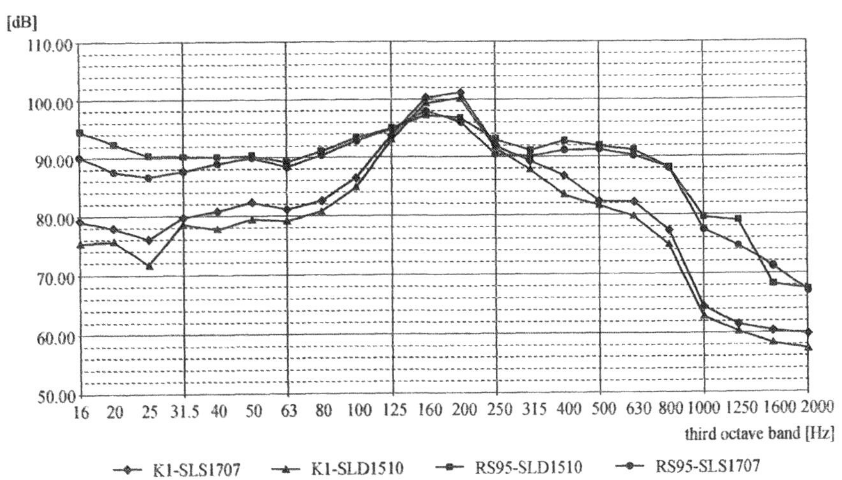

Figure 2.29 Results of measured velocities on different kinds of sleeper at the test track at Timelkam

results show that the impact on the K1 sleepers with rigid boots is higher than the impact on softer embedded sleepers in the ballast bed. This seems to be a plausible behaviour. For frame sleepers a completely different behaviour can be observed: the more rigid coating leads to a lower dynamic impact. Neglecting the minor influence of the rigidity of the coating, the large difference in the dynamic reaction of the two different kinds of sleeper is obvious.

If the dynamic impact factors given by the codes are realistic for classic concrete sleepers without any elastic layer it must be assumed that the influence of elastic elements on the stresses is not very high, whereas the influence of different sleeper geometries is enormous. The vibration velocities increase by a factor of about 1.5 for the frame sleepers in comparison with mono-block sleepers.

In summary, it is clearly shown that dynamic measurement and evaluation methods lead to realistic basic parameters for the design of sleepers and similar high dynamic loaded elements.

2.6.2 NOISE AND VIBRATION PROBLEMS

As well as load carrying, noise and vibration emissions of railway superstructures form one of the most important dynamic problems concerning railways [8]. Ambient vibration testing is a very valuable tool used to make these effects measurable and to find strategies to reduce these emissions.

The source of noise and vibration induction by trains is the contact surface of the wheels and tracks. This surface is not perfectly smooth and, owing to the moving train, vertical and horizontal impacts are permanently the cause of the vibrations. Starting on the rails, the vibrations propagate, for example, through the tunnel into the soil and to the foundation of buildings. This propagation follows the mechanical principles of wave propagation in soil.

To avoid negative effects (irritations, health defects) certain limits on permissible vibrations and also ground-borne noise in buildings situated near railway lines are established in national codes and regulations. The unit describing the vibrations is the KB value, which is a dimension-free value. It is defined as

$$1 \le f \le 2\text{Hz}, \qquad K_{\text{B}} = 28a \tag{2.3}$$

$$2 \le f \le 8\text{Hz}, \qquad K_{\text{B}} = \frac{33.5a}{f^{\frac{1}{4}}} \tag{2.4}$$

$$8 \le f \le 80\text{Hz}, \qquad K_{\text{B}} = \frac{160a}{f} \tag{2.5}$$

where
f = frequency (Hz)
a = acceleration (m/s^2)

The allowed limits depend on the global conditions (Table 2.5). If it is not possible to run a new railway line within these limits, measures for reduction of the vibrations are necessary. Otherwise it is impossible to obtain permission for construction and operation of a new railway line. The so-called 'insertion loss' is a unit describing the attenuation capacity of such measures. It gives the difference between the situations without a measure and the situation with a vibration attenuating measure [9].

To reduce vibrations different measures are possible both in the emission region in the subsoil and at the emission locations. The possibilities include improvements of the rolling stock, intensive maintenance of the roadway, modifications of the superstructure and improvements of the substructure. In addition, dynamic uncoupling measures for the affected buildings could also be practicable. Nevertheless, it would be sufficient simply to change the general layout of the line by moving it away from possible neighbours [10]. This last method may be an opportunity to avoid discussion, but owing to a number of other boundary conditions it is seldom possible.

For any kind of ground-borne noise and vibration reduction measure, vibration measurements are necessary in order to produce the design and the verification. An overview is now given of possible measures and therefore possible applications of vibration testing. On buildings several measures are possible. Without a claim to completeness some possibilities are (Figure 2.30):

Table 2.5 Limits for vibrations depending on the category of the environment according to ON S 9012

Category	Description	$\overline{K}_{B,S}$ for sufficient vibration protection		$\overline{K}_{B,S}$ for good vibration protection	
		Day	Night	Day	Night
1	Health resort, hospital	$\leq$6.0	$\leq$0.6	$\leq$3.0	$\leq$0.3
2	Residential area in suburb, weekend apartment area, rural residential area, schools	$\leq$8.0	$\leq$0.6	$\leq$4.0	$\leq$0.3
3	Urban residential area, area of agriculture and forestry with apartments	$\leq$8.0	$\leq$0.6	$\leq$4.0	$\leq$0.3
4	Centre, area of companies without noise pollution	$\leq$10.0	$\leq$0.8	$\leq$6.0	$\leq$0.4
5	Area of companies with small noise pollution	$\leq$12.0	$\leq$12.0	$\leq$8.0	$\leq$8.0

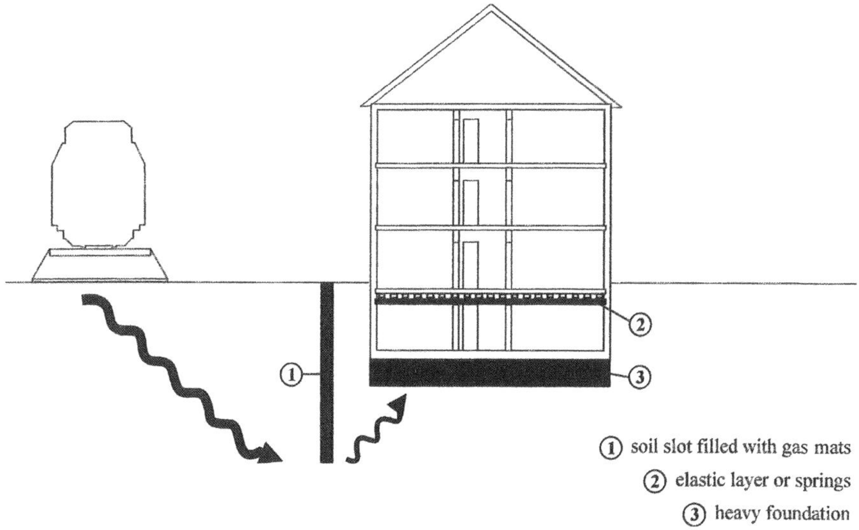

Figure 2.30 Examples of measures on buildings

- elastic elements in the foundation (uncoupling the building from the ground supporting medium);
- heavy foundations (high impedance);
- soil slot stabilized with gas-filled mats near the building (interruption of wave propagation);
- natural frequencies of slabs and walls out of the range of main frequencies of vibrations induced by the trains (avoiding resonance effects).

The main measures focus on the impedance of the tunnel, which means that high impedance reduces the vibrations of the tunnel shell, thus reducing the vibrations propagated through the soil to the buildings. High impedance can be reached by optimizing the cross-section of the tunnel. For example, round cross-sections are more rigid than rectangular cross-sections. The rigidity of the tunnel shell is also influenced by the construction principle. Bored tunnels with pre-cast tunnel sections are very elastic; these shells behave like a chain (short elements connected by hinges). To increase the impedance it is also possible to construct thick and heavy tunnel floors or to construct double-shell tunnel tubes (Figure 2.31).

Measures on the track itself can vary. For conventional ballasted superstructures elastic layers uncouple the track from the substructure and lead to a decrease of the vibrations induced into the ground. Such elastic layers might be sub-ballast mats or booted sleepers (under sleeper pads), as explained in Figure 2.32.

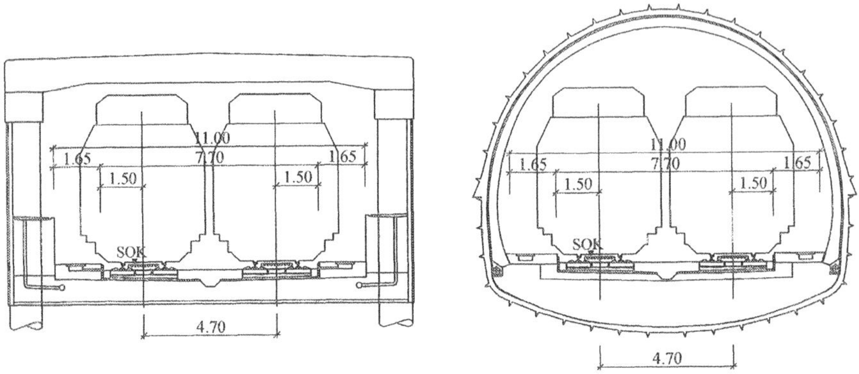

Figure 2.31 Different cross-sections for tunnels (all dimensions in m)

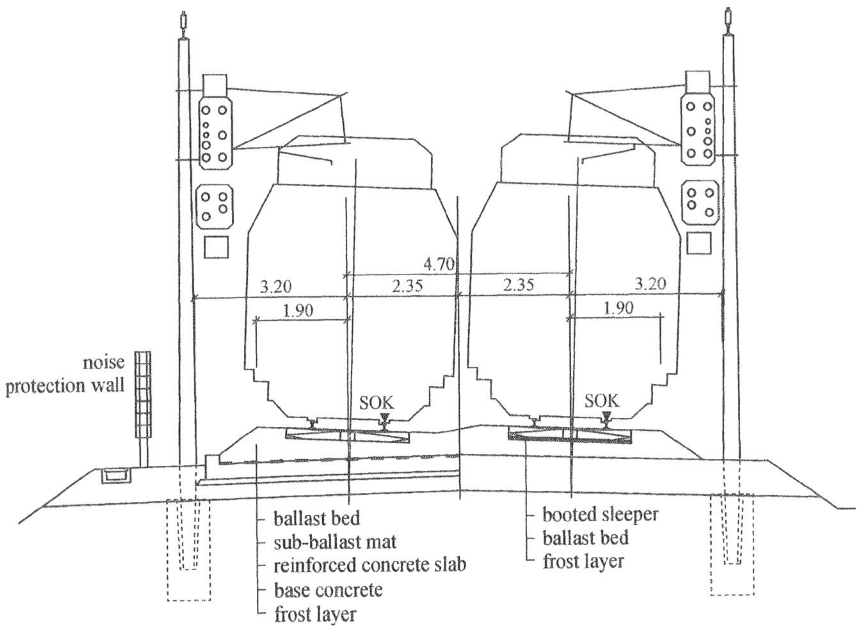

Figure 2.32 Ballasted track with sub-ballast mats or booted sleepers (under sleeper pads) (all dimensions in m) [11]

Of course it is also possible to increase the elasticity of the rail pads and/or base-plate pads (Figure 2.33), but the variation of the stiffness of these elements can only be varied within a small bandwidth. Too low rigidity of these elements negatively affects the run of the rolling stock, leading to an increase in the

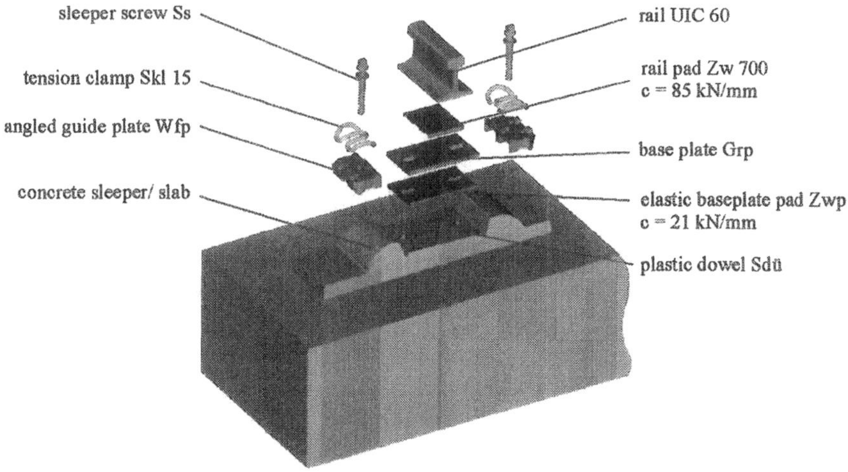

Figure 2.33 Rail fastener with a rail pad and base-plate pad according to Vossloh

stresses in the rails and deflections of the rails. Furthermore, air-borne noise may increases owing to possible greater movements of the rails.

For ballast-less systems more effective noise and vibration attenuating modifications are possible: the so-called 'floating track slabs' or classical mass–spring systems (Figure 2.34). For such systems many construction principles have been developed within the last 20 years [12].

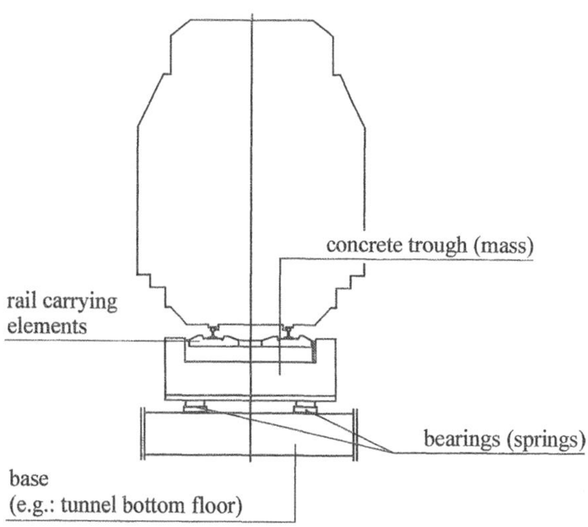

Figure 2.34 Principle cross-section of the floating track slab system

According to the mass and natural frequency, a division into the following three groups of mass–spring system (MSS) is made:

- lightweight MSS with $m \leq 4\,\text{t/m}, f_1 \geq 15\,\text{Hz}$;
- mediumweight MSS with $m \leq 8\,\text{t/m}, f_1 \geq 10\,\text{Hz}$;
- heavyweight MSS with $m > 8\,\text{t/m}, f_1 < 10\,\text{Hz}$.

All of these measures have the same mechanical background. This basic principle can be explained by a linear single-degree-of-freedom system consisting of a mass, a spring and a damper. Figure 2.35 shows that dynamic forces applied on such a system lead to increased dynamic forces on the foundation for frequency ratios below $2^{\frac{1}{2}}$. For higher frequency ratios reduction of the dynamic forces on the foundation can be observed.

It is therefore clear that the natural frequencies and damping ratios of all these systems are essential for the vibration attenuating capability. It is necessary to measure these values on recognized systems for evaluation, as well as a proof of construction quality and also for maintenance reasons. Ambient vibration testing is one of the simplest methods of doing this. In the past a lot of testing indicated that the results of ambient testing fit very well with the results of forced vibration testing.

The following examples give some indication of the standard for noise and vibration reduction used by the Austrian Federal Railway Company ÖBB and the application of ambient vibration testing to these superstructures. The examples are the result of extensive investigations into research and development of noise and vibration attenuating track constructions in the last ten years [13].

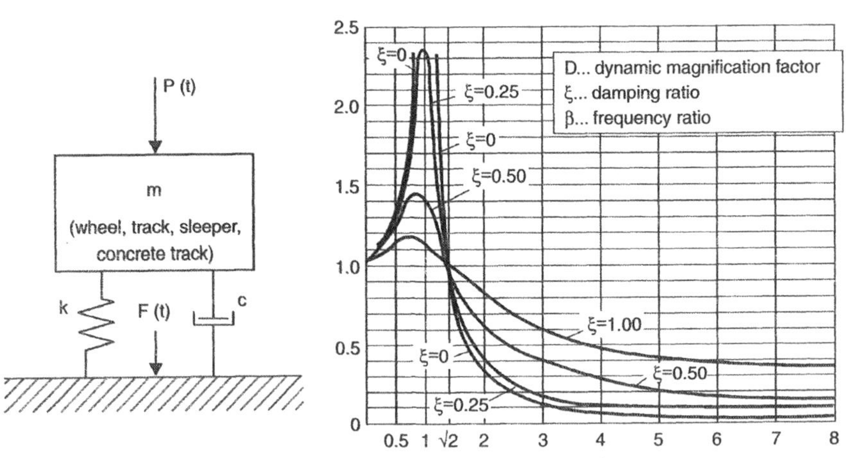

Figure 2.35 Mechanical principle of mass–spring systems

Continuing with the example above using different kinds of sleepers and different levels of sleeper vibrations, the consequences to vibration propagation in the soil are shown in Figures 2.36 and 2.37. The measured results show that the initial differences of vibration levels on the sleepers affect the vibration propagation through the soil. Furthermore, it can be seen that the mechanic principles of wave propagation lead to a reduction in the difference between the vibration levels for the different kinds of sleeper and under-sleeper pads. This shows the importance of using measurements to obtain knowledge of the vibration attenuating effects of different kinds of modification of superstructures.

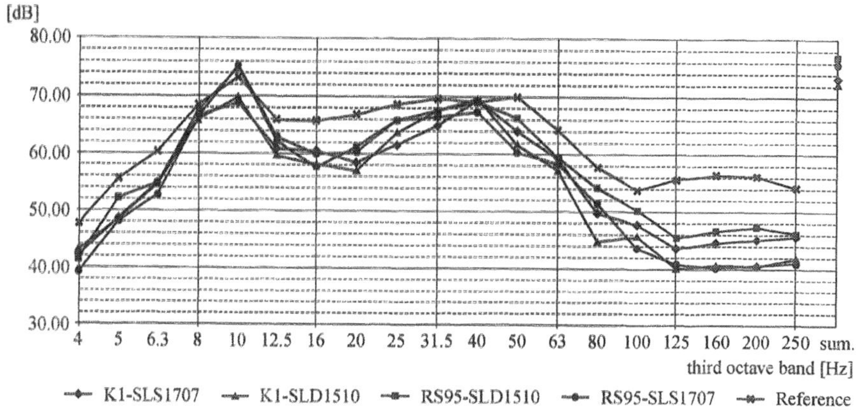

Figure 2.36 Third-octave velocity level measured on the ground at a distance of 10 m from the track

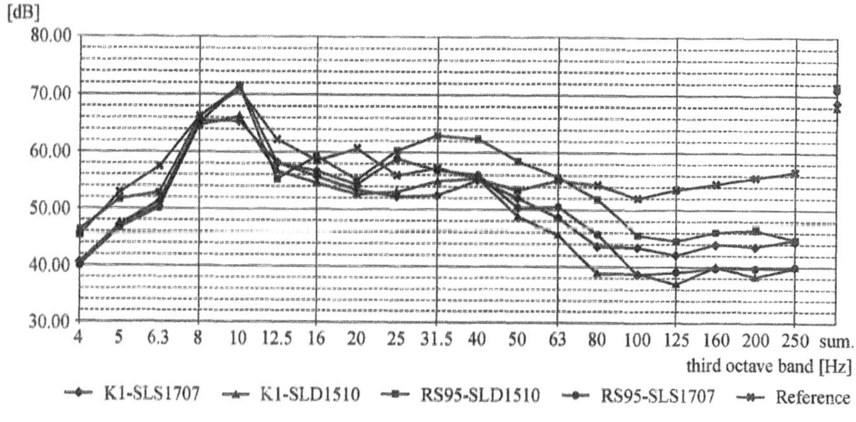

Figure 2.37 Third-octave velocity level measured on the ground at a distance of 20 m from the track

Possible optimizations of insertion loss of under-sleeper pads were tested on a test track near Riedau in Upper Austria. The design criteria for the elastic elements (sub-ballast mats and under-sleeper pads) were to obtain the same deformations for different kinds of superstructure. Figure 2.38 shows the results for the insertion loss. The reference superstructure is a conventional ballasted track with pre-stressed concrete sleepers and rail type 60 E1.

The results show that all three modifications of the superstructure resulted in similar insertion losses. This is in accordance with theory, owing to the fact that the natural frequency of all three systems is the same. In this case ambient vibration testing was used to prove the theory and the design of the elastic elements.

As another example, the mass–spring system in the Römerberg Tunnel shows what advantage may be generated by using ambient vibration testing to monitor the time-dependent behaviour of ballast-less floating track slab systems. Situated in Upper Austria, the Römerberg Tunnel is part of the new Westbahn railway line and was finished in 1995. As some residential buildings were situated above the tunnel at a distance to the tunnel shell of only about 10–20 m, extraordinary vibrations and ground-borne noise were forecast for these buildings. Therefore it was decided to modify the track construction by putting it on a floating track slab. Figure 2.39 shows the basic alignment of the mass–spring systems in the tunnel [14,15].

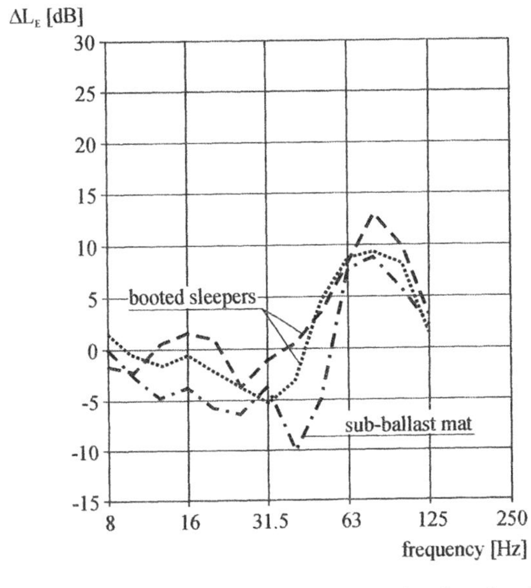

Figure 2.38 Results of a test track near Riedau: insertion loss for different systems

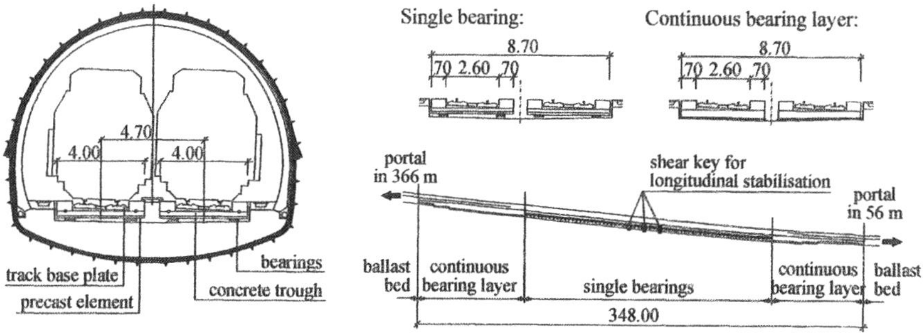

Figure 2.39 Mass–spring system in the Römerberg Tunnel (all dimensions in m)

To reach the limits for good noise and vibration protection in residential areas according to ON S 9012 a medium-weight mass–spring system had to be used. The result of the prediction model, based on linearity for both main construction elements, bearings and reinforced concrete trough, was that a natural frequency of approximately 13 Hz and a mass of 6 t/m would be sufficient. Knowing these parameters, a joint-less floating slab system was developed and put into practice.

The system has a total length of 348 m, 192 m being supported on single bearings with the remaining parts of the slab lying on a continuous bearing layer. For horizontal stabilization of the floating reinforced concrete trough shear keys with vertical elastic bearings were developed. At both ends of the MSS, instead of rail expansion joint constructions, a continuous connection between the MSS and the conventional ballast bed superstructure was designed.

During the construction of and since finishing the MSS in the Römerberg Tunnel many different measurement programmes have been carried out. The aims of these investigations are to gain knowledge of the static and dynamic behaviour of the MSS under real conditions.

After finishing the trough construction, VibroScan® tests were made to ensure that all components of the system were operating correctly. These tests showed that the prediction model for the vibration propagation through the soil and for the insertion loss of the MSS was accurate. Ambient vibration measurements and evaluations with the dynamic measurement and testing system BRIMOS® showed that the natural frequencies and mode shapes fit very well to the calculated ones for the not cracked MSS and the 40 Hz bearing stiffness. Figure 2.40 shows a comparison between two measurement results of the VibroScan® tests and the predicted insertion loss [10,16,17].

The comparison shows that up to 32 Hz the measured results are identical to the calculated ones; for higher frequencies the calculated insertion loss is greater than that actually measured. The reason for this is not an insufficient

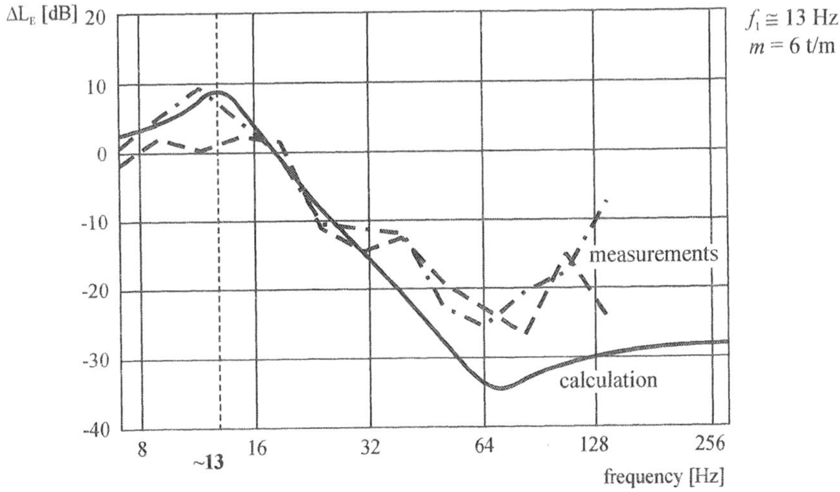

Figure 2.40 Insertion loss of the MSS in the Römerberg Tunnel [16]

prediction model but the fact that the VibroScan® tests were made on a pure concrete trough without the mass of the rails and the rail carrying elements [12]. Nevertheless, the spectral fast Fourier transformation (FFT) analysis of the VibroScan® sweeps and the BRIMOS® results fit very well together. Both investigations led to a highest effective vertical natural frequency of approximately 11.8 Hz for the single bearing section (Figure 2.41). In the year 2002, measurements of the dynamic characteristic showed only very minor differences from the results of the first measurements in 1996.

For this application ambient vibration testing is a very effective tool to use for quality control and for maintenance. Therefore the technology is used for

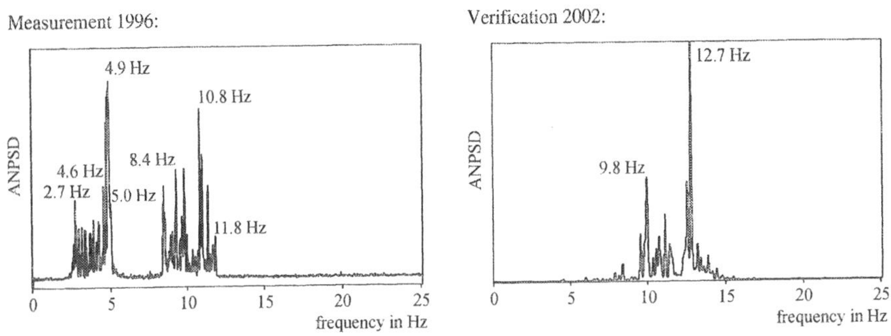

Figure 2.41 Response spectra of the floating track slab system in the Römerberg Tunnel

all new floating track slabs as an acceptance test and for regular inspections during operation.

2.7 LIMITATIONS

2.7.1 *LIMITS OF MEASURING TECHNOLOGY*

The dynamic measurements in the scope of the AVM show different claims to measuring accuracy. Being dependent on the desired result, measurement errors play a more or less important role. There is a distinction between:

- measurements of the dynamic characteristic of a structure where only relative values are determined and
- measurements of the actual dynamic behaviour of structures where absolute values for deformations are measured.

If no absolute values are required for the amplitudes, measurement errors play a secondary role. It cannot be expected that the measured value exactly represents the measurable variable in dynamic measurements. It must be assumed that the measuring result is distorted by external influences, shortcomings in the measuring chain and the measuring procedure used. Furthermore, subjective influences are to be expected by deployment of the measuring staff. These facts show that the final purpose of the measurement should be known before and should be included in test planning.

The difference δ_{absolute} between the measurable variable u_{real} and the measured value u_{measured} is called the absolute measurement error:

$$\delta_{\text{absolute}} = u_{\text{measured}} - u_{\text{real}} \tag{2.6}$$

The relative measurement error is explained as

$$\delta_{\text{real}} = \frac{\delta_{\text{absolute}}}{u_{\text{real}}} \tag{2.7}$$

and is often specified as a percentage. As the accurate value of u_{real} is not known, the measured value u_{measured} is often used as the reference value.

The proportional error for analogous electrical measuring instruments mostly refers to the end scale value. The relative error is considerably greater for small amplitudes and therefore an attempt should be made to place the measured values in the upper third of the scale range. In the present case this is reached by situating the signals in a conditioned way, i.e. in the optimum measurement range.

The amplifiers used are chosen in such a way that the amplification is done with a potential of two and not with the usual potential of ten. Therefore a fine adjustment is easier. An absolutely correct representation of the measured value is deliberately avoided. Determination of the actual measured values must be done by means of calibration, which occurs via measurements with a method (laser) applied in parallel.

Essentially three types of measurement errors can be distinguished:

- Systematic measurement errors, such as temperature influences and non-linear phenomena, pre-stressing or other external influences, can be corrected by an adequate calibration.
- Random errors – as, for example, not identified disturbances by other electrical applications – can be discerned and their size estimated by repeating the measurement.
- Gross measurement errors, e.g. reading errors, misunderstandings, loose contacts, equipment damages or similar incidents, are basically avoidable and must be prevented by good training and organization of the measuring team.

It is essential to keep measurement errors as small as possible because the expenditure for a correction can become unjustifiably high or realistic corrections are often not possible.

When time-dependent values are measured, the frequency response of the measuring equipment and their elements play a decisive role. Provided that the time dependence of the measurable variable is sinusoidal, the systematic error due to the frequency response while measuring frequency can be easily corrected. Here the wrong application of filters in the measuring chain must be particularly pointed out. These filters can remove essential parts from the signal.

If in a measuring signal frequencies are contained where the frequency response of the measuring system deviates from a constant value, considerable errors occur, which can only be corrected in the scope of a frequency analysis. The most frequent dynamic measurement errors in the described system are as follows:

- The sensors used are too sensitive and cannot be brought into a zero position before the beginning of the measurements. This results in an offset in the measurement, which is reflected in the data record. When it is eliminated, however, valuable information can be lost and therefore the result can be distorted. The use of high-pass filters must not be generalized because in the case of very big bridges it is easy to fall into the range of the first eigenfrequency and thus the basic information becomes useless. Therefore the aim is to record the unfiltered signal and to solve the problem via eligible digital filters during evaluation.
- Practical measurements taken since 1996 have shown that disturbances from electrical interactions sometimes appear in the signals. However, bridge

frequencies over 20 Hz are interesting in a few cases only and so higher frequencies can be filtered out without any loss of information. More far-reaching considerations showed that this should already be done in an analogous stage. Therefore low-pass filters, which can be individually adjusted, are installed in the measuring chain. Part of test planning should be to estimate and determine this filter setting before a measurement is taken.

- The most substantial disturbances and therefore source of errors during ambient measurement are forced vibrations. Above all, the applied clearly defined vibrations and the so-called 'combination' frequencies, which are developed in the range of a dying down forced signal, produce trouble. This error can be remedied by cutting out only those ranges from the long recording series that do not show any disturbances. If this is not possible, disturbances have to be reduced by means of algorithms or be marked as disturbances and interpreted in the report. Thanks to the experience of the personnel such frequencies can, however, always be excluded from the assessment.

The manufacturers of measuring instruments state the frequency response and the errors resulting from the individual dispersion of the frequency response curves for measuring systems of dynamic values. Naturally the errors to be considered are larger at the edges of the frequency responses. Whereas the errors in the main employment area are at a magnitude of 1%, in the vicinity of the critical frequencies errors of up to 10% are found. Such errors are tolerable for qualitative evaluation, but for quantitative assessments they are problematic. Therefore it is imperative that data records at the limit frequency are eliminated.

A further source of errors during quantitative evaluation is the sensitivity of the system to differences in the global and local systems of coordinates. The sensors can only be installed in such a way that their local system of coordinates complies with the global system of the structure. Vibration of the structures generally represents a three-dimensional movement. A point where the measurable variable is received moves on an ellipsoid in the case of harmonic movement. In the case of three-dimensional sensors the movement description is sufficient to be able to carry out a transformation of coordinates. In the case of one-dimensional sensors it is measured only in one direction. The movement components vertical to this direction distort the measuring result considerably in the event of small movements in the measuring direction and large movements vertically to them.

2.7.2 LIMITS OF APPLICATION

As already mentioned, the AVM serves the registration of structural and mechanical behaviour of structures. Accordingly damages that have no effect on the vibration behaviour cannot be determined. Examples for this are mentioned without any claim to completeness:

- damaged guardrails, handrails, etc.;
- advanced carbonization front;
- beginning reinforcement corrosion, which has not yet led to an essential weakening of the steel cross-section.

On the other hand, it must be mentioned in this connection that visually apparent damage (e.g. guardrails, concrete spalling) is also determined in the course of work with the AVM. Damage that has no effect on the load-bearing behaviour at first sight is frequently recognized. For example, damage of the pavement often leads to increased acceleration of the structure during the passage of cars, which can be recorded very well using the AVM. Similar processes can be determined in the case of defective road joints.

2.7.3 LIMITS OF ANALYSIS

Engineering buildings are usually complex structures exposed to varied, often unknown, stress and environmental influences. Every building is individually established and therefore full of imponderables in its load-bearing behaviour. It is therefore necessary to choose the measuring locations and condition observations on the basis of all available information on the foundation soil, construction progress, building materials, construction firms, construction condition, load history and possible existing damage with engineering reason. As not all structure can be globally provided with sensors, it is possible that not all relevant stress and damage effects can be recognized.

The evaluation and interpretation of monitoring results make high demands on the interdisciplinary knowledge and experience of the monitoring personnel. With increasing application of such monitoring systems, generally accessible and improved findings on significant influence values and their evaluation are gained and wrong interpretations are avoided. The composition of a data bank, which includes all relevant data for the structure to be monitored, supports and accelerates this process.

The monitoring of technical systems is based on knowledge of the structural properties of the system to be observed. Only if the system behaviour is sufficiently known can the monitoring sensors be arranged at the correct points and an evaluation made of the measurable variables with regard to relevant core values for system identification. For localization of the areas of the structure with the highest stress both calculation and measuring methods are suitable.

It must be checked in every case which method is possible and can be efficiently applied for the structure to be monitored. Thus in the run-up to the measurements important decisions have to be taken, which influence the quality of the results. According to the monitoring order certain physical values have to be given priority treatment. In the case of the method developed here vibration measurements are carried out and consequently their frequencies of resonance and the respective vibration forms are determined. The change in specified parameters can be regarded as an indicator of structural changes.

2.7.4 PERSPECTIVES

The development potential of the method is huge, in particular with regard to the increasingly improving possibilities of measuring technology and computer capacity. Accordingly such systems are being developed worldwide. For final acceptance of the method it will be necessary to intensify the training in issues of construction dynamics and integrate respective information on practical problems into the theory. Acceptance will be gained if the application of expert systems and probabilistic considerations are given more attention in structural assessment. By means of approaches using fuzzy logic it will be possible to identify damages automatically. This will, however, not replace the judgement of the bridge engineer in the foreseeable future. Therefore it seems to be very important that the method is operated by bridge engineers who have acquired knowledge of measuring technology, particularly in the years of its development. For the engineer this is a tool for the quantification of several subjectively noticeable phenomena.

The desired targets for development in the near future are:

- the development of a monitoring system without cables;
- the development of faster algorithms for the evaluation of measurements;
- the formulation of damage scenarios and algorithms for the automatic recognition of problems;
- the integration of the method in expert systems.

Despite the high potential of the method it must be pointed out that traditional methods cannot be replaced but should only be complemented. This enables further development of traditional methods. A far-reaching combination between visual inspection, conventional inspection methods and dynamic vibration analysis is advisable.

Due to the fact that measuring technology already exceeds the requirements that can be utilized within a calculation, it is useful to gather data in order to receive data series covering several years. Even if today no 100% valid conclusion can be drawn, such data will be particularly significant for a later assessment. For this reason measurements are carried out for every newly built bridge in Switzerland.

REFERENCES

1. Eibl, J. *et al.* (1988) Baudynamik, in *Betonkalender 1988*, Band II, Ernst & Sohn, Berlin.
2. Cantieni, R. *et al.* (1996) *Untersuchung des Schwingungsverhaltens großer Bauwerke*, Technische Akademie, Esslingen.
3. Wenzel, H., Pichler, D. and Schedler, R. (1999) *Ambiente Schwingungsmessungen zur System- und Schadenserkennung an Tragwerken*, Bauingenieur 74, Heft 3, pp. 115–23.
4. Frangopol, D.M. and Das, P.C. (1999) Management of bridge stocks based on future reliability and maintenance costs, in *Bridge Design, Construction, and Maintenance*, Institution of Civil Engineers, Thomas Telford, London, pp. 45–58.
5. Wenzel, H., Geier, R. and Eichinger, E. (2001) *Untersuchungen anlässlich des Abbruches ausgewählter Tragwerke*, Endbericht in Kooperation von VCE und Technical University Wien, Wien, October.
6. Esveld, C. (2001) *Modern Railway Track*, 2nd edn, Delft University of Technology.
7. Rutishauser, G. and Pichler, D. (2001) *Masse-Feder-Systeme: Erfahrungen und Stand der Technik*. Proceedings of the Getzner-Congress, 17–18 May 2001, Brand near Bludenz.
8. Rutishauser, G., Steinhauser, P., Honeger, Ch., Flesch, R., Kalivoda, M.T., Hasslinger, H.L., Schilder, R. and Pichler, D. (2003) *LEO – Low Noise and Low Vibration Track*. Proceedings of the LEO Seminar, 6–7 March 2003, Vienna.
9. Pichler, D. (2003) *Vibration Attenuating Measures for Railway Lines – Floating Track Slab Systems*. Conference Proceedings of the Rail-Tech Europe 2003, Vol. 1, bis 3, April 2003, Utrecht, pp. 89–98.
10. Wenzel, H., Pichler, D. and Rutishauser, R. (1997) *Reduktion von Lärm und Vibrationen durch Masse-Feder-Systeme für Hochleistungseisenbahnen*. Oral presentation at the D-A-CH-Meeting in Zürich, SIA-Dokumentation D 0145, pp. 123–32.
11. Schilder, R. (2004) *USP – Under Sleeper Pads*. Proceedings of the ÖVG Tagung, 14–16 September 2004, Salzburg.
12. Pichler, D. and Zindler, R. (1999) *Development of Artificial Elastomers and Application to Vibration Attenuating Measures for Modern Railway Superstructures*. Proceedings of the First European Conference on Constitutive Models for Rubber, 9–10 September 1999, Vienna, Austria, A.A. Balkema, Rotterdam, Brookfield, pp. 257–66.
13. Pichler, D., Mechtler, R. and Plank, R. (1997) *Entwicklung eines neuartigen Masse-Feder-Systems zur Vibrationsverminderung bei Eisenbahntunnels*, Bauingenieur 72, Springer-VDI-Verlag, Berlin, pp. 515–21.
14. Pichler, D. (1998) *Concrete Based Floating Track Slab Systems – Modelling and Reality*. Proceedings of the EURO-C 1998 Conference on Computational Modelling of Concrete Structures, 31 March–3 April 1998, Badgastein, Austria, A.A. Balkema, Rotterdam, Brookfield, pp. 665–71.
15. Pichler, D. and Huber, P. (1997) *Reduction Measures for Tunnel Lines*, Report for RENVIB II Phase 1 to ERRI, Vienna Consulting Engineers and Rutishauser Ingenieurbüro.
16. Steinhauser, P. (1996) *Römerbergtunnel – Ergebnisse der VibroScan® Untersuchung zur immissionsmäßigen Abstimmung des Oberbaus*. Report to HL-AG, Vienna.
17. Steinhauser, P. (1997) *Römerbergtunnel – Ergebnisse der VibroScan® Untersuchung auf dem Masse-Feder-System*. Report to HL-AG, Vienna.

3

Feedback from Monitoring to Bridge Design

Structural health monitoring has become widely accepted in bridge management. The methodologies have been considerably developed over the past ten years and have reached a certain maturity. Many lessons have been learned from the monitoring activities. This paper highlights the lessons learned over time, which have relevance to bridge design.

3.1 ECONOMIC BACKGROUND

The transportation infrastructure is ageing. Bridges are an essential asset of the economy. Reference is made here to the situation in the United States, where within the network of the Federal Highway Agency (FHWA) 590 000 bridges are serving, out of which 160 000 are rated as deficient when traditional methodology is applied. Replacement costs of US$7 billion annually over 20 years are estimated to achieve a perfect upgrade. In order to avoid such costly situations, the lessons learned with relevance for design should be considered in present bridge design processes. Another drastic example is a bridge built in Austria in 1978, following the minimization principle of construction costs, at €8.5 million. Within 25 years a total of €19.5 million had to be invested into retrofit measures. With consideration of life cycle cost approaches such situations have to be avoided in the future.

Ambient Vibration Monitoring H. Wenzel and D. Pichler

3.2 LESSONS LEARNED

The following is a collection of lessons learned from monitoring of over 400 bridge structures since 1997. The structures monitored are mainly situated in central Europe and represent the typical design for this region. Nevertheless, most of these lessons can be extended to structures worldwide.

3.2.1 CONSERVATIVE DESIGN

Monitoring of over 400 bridges clearly shows different behaviours of structures designed following different philosophies. Bridges designed conservatively are not affected by dynamic phenomena that generate concern or trigger damage. Bridges with the design focused on economy very often do not have any reserves to cover the extraordinary loadings that appear in reality. The difference in dynamics becomes obvious in Figures 3.1 and 3.2. Conservative bridges show a high system damping with distinct characteristics. Economic designs very often come close to areas where resonance appears. This resonance might have a very local and limited effect, but over time it leads to damage in structures.

From bridge management we know that an additional investment in 10% higher quality makes a difference of over 200% in costs over the life cycle of a bridge of 100 years. Drastic examples experienced are a bridge composed of single span I-girders, designed to the limit, which has consumed 220% of the investment costs in retrofit over a period of 25 years. The opposite is the resistance of a duly designed box girder bridge that survives a displacement of a single pier of 110 cm that could be retrofitted at reasonable costs.

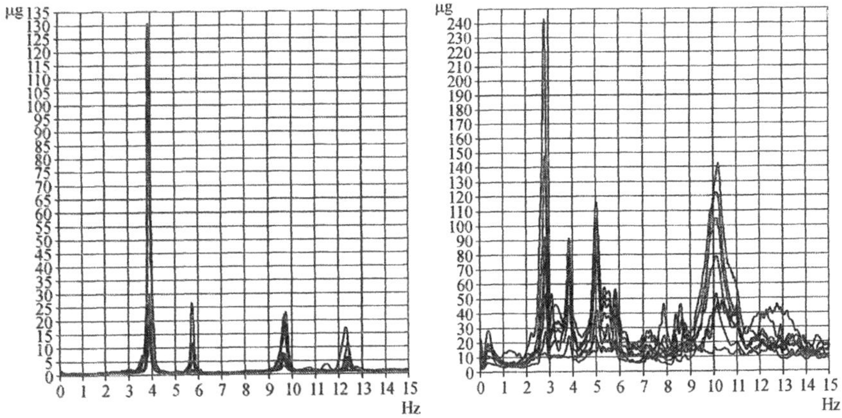

Figure 3.1 Spectrum of a sound bridge (left) and spectrum of a damaged bridge (right)

Figure 3.2 Resistant box girder (left) and costly I-girders (right)

3.2.2 EXTERNAL VERSUS INTERNAL PRE-STRESSING

Damage found on grouted internal cables triggered a dramatic change of design philosophy. Some countries, such as Germany, specified that new bridges have to be built using external cables only. This is for the purpose of inspectability and eventual replacement. The experience made while testing 30 bridges built in the late 1950s and early 1960s showed that in only one of the structures was damage of the cables found. In all other bridges, where damages were suspected, no evidence of corrosion or wire breaks has been detected. The only damaged bridge was a long way from malfunction. On the other hand, the bridges with external pre-stressing often show cracks in the anchoring parts and are often unequally stressed. The best results have been received on bridges where grouted tendons embedded in concrete in combination with external cables have been chosen.

3.2.3 INFLUENCE OF TEMPERATURE

The design codes for bridges provide clear instructions on how to consider the temperature effect in bridge design. These instructions normally give a high and a low temperature to be considered and eventually a temperature gradient between the bottom and top of a structure. No reference is made to the type of bridge or the material used. Monitoring provides the chance to record the actual effects of temperature on structures exactly. The lessons learned are actually easy to accept:

- Slender structures react very close to the provisions of the codes.
- Stiff structures very often deviate considerably from the expected stress distribution.
- The temperature gradients actually recorded on stiff structures by far exceed the values of the design.

- Temperature load cases can be the decisive load cases.
- Temperature changes do not trigger a linear behaviour. Below 5 °C a clear stiffening effect is recorded.

The stiffness of concrete bridges depends on temperature. The relation is given in Figure 3.3, which shows an almost bilinear condition measured on a classical post-tensioned concrete box girder bridge. This has to be considered in the interpretation of monitoring results.

Steel bridges show quick reactions on changing temperatures. The records of a 5 m high steel box girder are shown in Figure 3.4. There is a difference between heating or cooling periods. The sensors represent the outside temperature and the inside temperature on the bottom slab and on the deck slab. The patterns shown here are representative.

Temperature changes in the annual cycle are homogeneous. Figure 3.5 shows the behaviour of a concrete mass supported on bridge bearings in a railway tunnel. There is no influence of sunshine. Temperature conditions of the surrounding soil are quite stable, but the trains transport air from outside through the tunnel. The graph shows how the structure behaves homogeneously over the years. The maxima and minima values measured are actually higher than expected. The results for a concrete bridge (refer to Figure 3.6), where concrete is used in the shape of a box girder with cantilever arms, also shows a homogeneous cycle. The maximum and minimum temperatures according to the codes are never reached. The structure reacts moderately to warming or cooling. No clear daily cycle can be isolated.

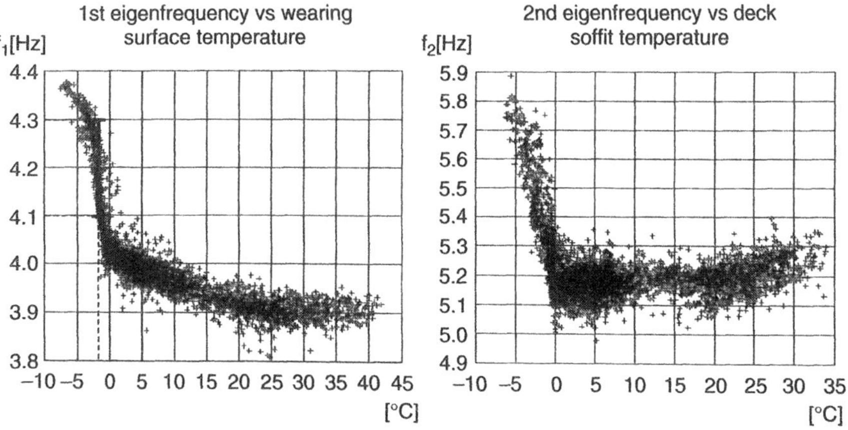

Figure 3.3 First eigenfrequency versus wearing surface temperature and second eigenfrequency versus deck soffit temperature [1]

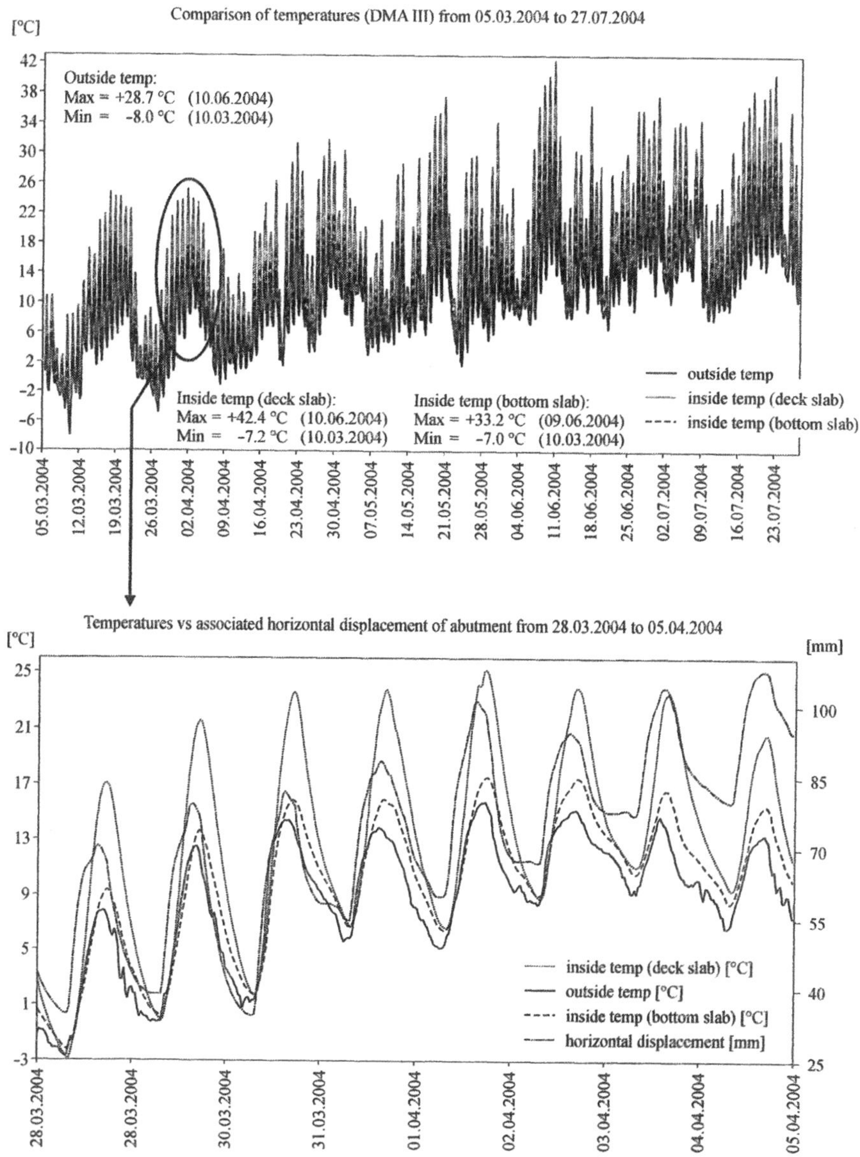

Figure 3.4 Representative temperature sensor records and longitudinal displacement of a steel bridge abutment (DMA is German abbreviation for permanent measurement system)

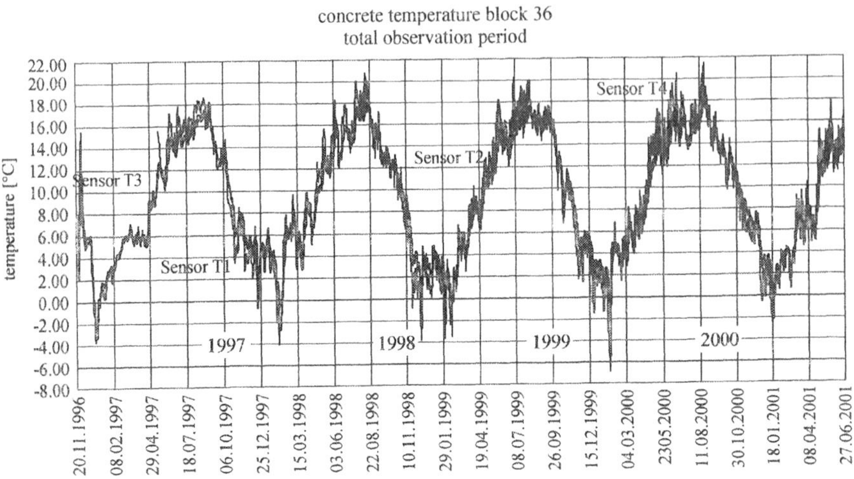

Figure 3.5 Long-run behaviour of a concrete mass supported on bridge bearings in a railway tunnel

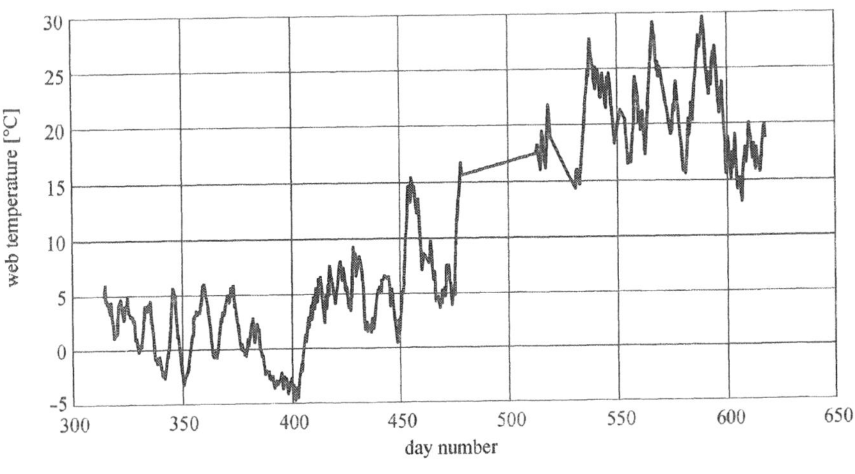

Figure 3.6 Variation of web temperature of the z-24 bridge observed over a period of one year [2]

A typical steel structure shows a rather violent reaction on temperature changes (Figure 3.7). The reaction is quite quick and produces strain in the system. This might be because of bearing friction, which is released suddenly causing a displacement of the structure. This effect can be particularly harmful to the expansion joints, and also to the bearing. Particular bridges that are

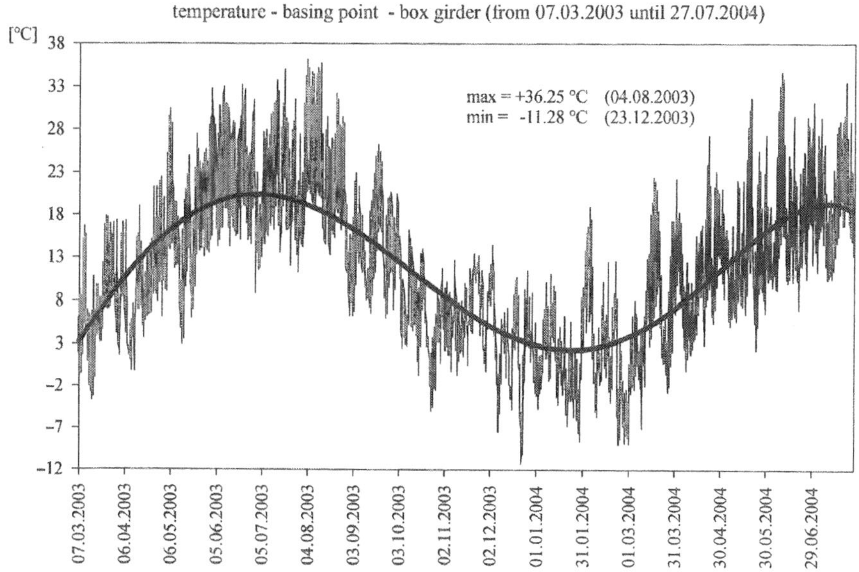

Figure 3.7 Temperature conditions inside the steel box girder at the Europa Bridge on the Brenner Motorway

not straight in plan might introduce exceptional forces into weak axes of the outfitting.

The consequence of these experiences should be an individual application of temperature loads depending on the type of structures and conditions they have to bear. The following implications might be considered:

- to increase the loads from differential temperatures in stiff structures;
- to increase the temperature range considered in steel structures;
- to look at the effects of quick temperature changes on the global behaviour.

3.2.4 *DISPLACEMENT*

Bridges are flexible and displace under various loads. Displacements are calculated using structural models or finite element calculations. These models very often do not reflect reality. A typical case is the displacement of a certain steel bridge due to temperature changes, which are shown in Figure 3.8, while monitoring results provided displacements according to Figures 3.9 and 3.10. The latter demonstrates the fact that the displacements of this steel bridge abutment are remarkably higher than those obtained from theoretical, linear

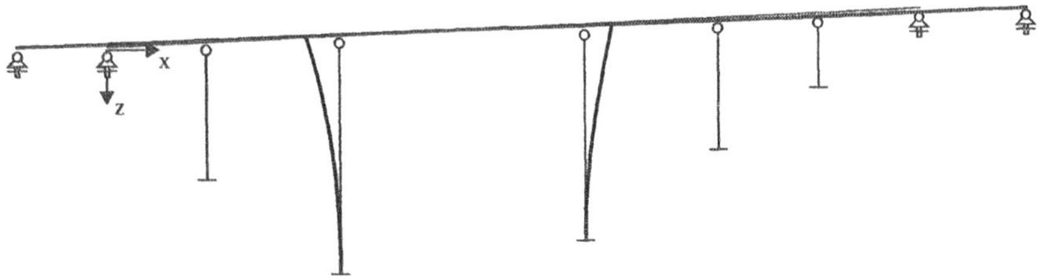

Figure 3.8 Displacement of a bridge due to temperature changes affecting only the superstructure

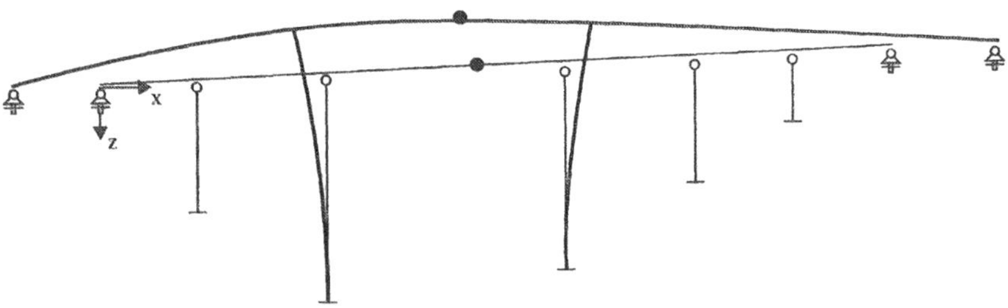

Figure 3.9 Displacement of a bridge due to temperature changes recorded by monitoring

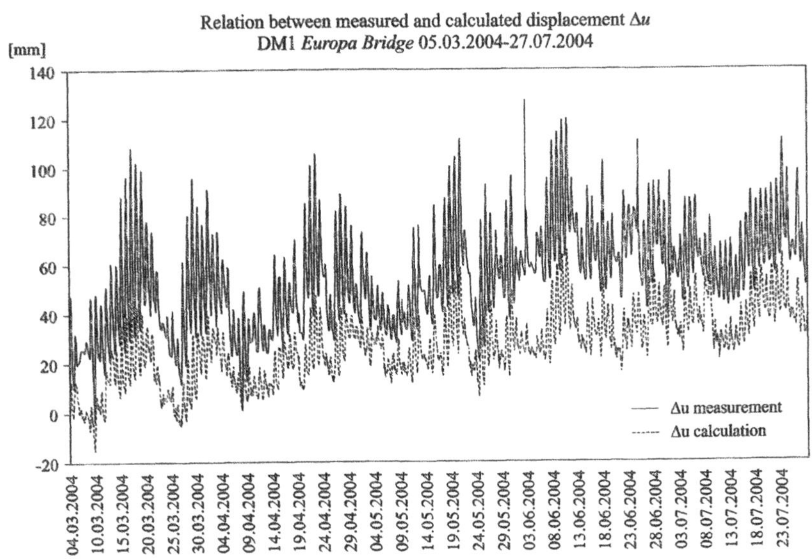

Figure 3.10 Comparison between measured and FE (finite element)-based displacements of a steel bridge abutment

elastic calculations (uniformly distributed temperature). The difference is mainly related to the following facts:

- The prevailing temperature gradients affecting the bridge piers contribute vastly to the total horizontal displacement.
- The stiffness of the columns very much depends on the degree of fixation of the pier in the foundation.
- Bearings do not show a linear behaviour all the times and tend to be stiff until a certain minimum force has been reached.
- A certain stress limit has to be reached before restoring forces are activated, particularly when elastomeric bearings are provided.

In major bridges sudden displacements of ±50 mm have been recorded, which can be attributed to suddenly released restoring forces of bearings. This displacement is normally within the regular limits of allowable displacement, but the sudden reaction might trigger secondary problems, such as restraints in the expansion joint. A number of failures of expansion joints have been attributed to these phenomena. The frequency of such phenomena is not sufficiently documented yet. In the record of a major steel bridge of only six months, three such occasions have been detected (Figures 3.11 and 3.12). The sudden displacement is also visible in a signal of three-dimensional accelerators which is placed at the top of a pier (Figure 3.13). A displacement due to bearing

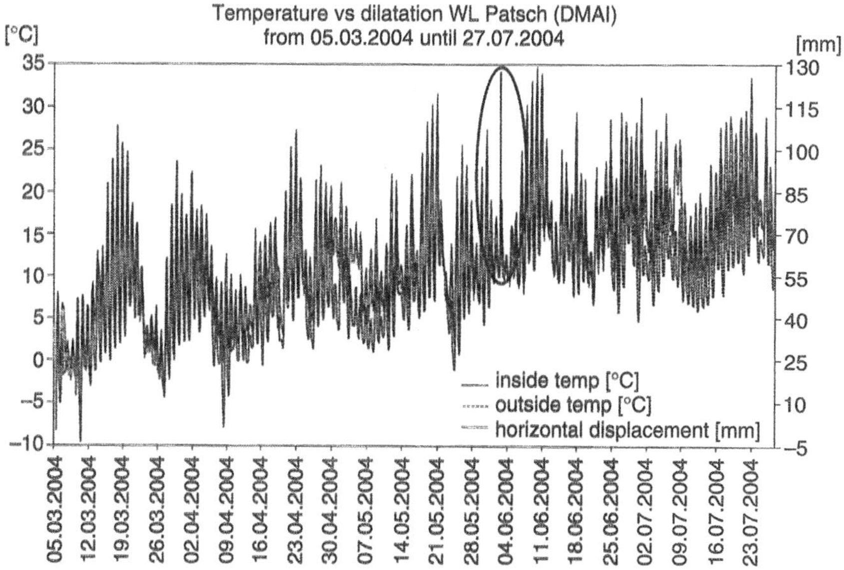

Figure 3.11 Uncommon, sudden reactions in the displacement recordings of the abutment over a period of five months

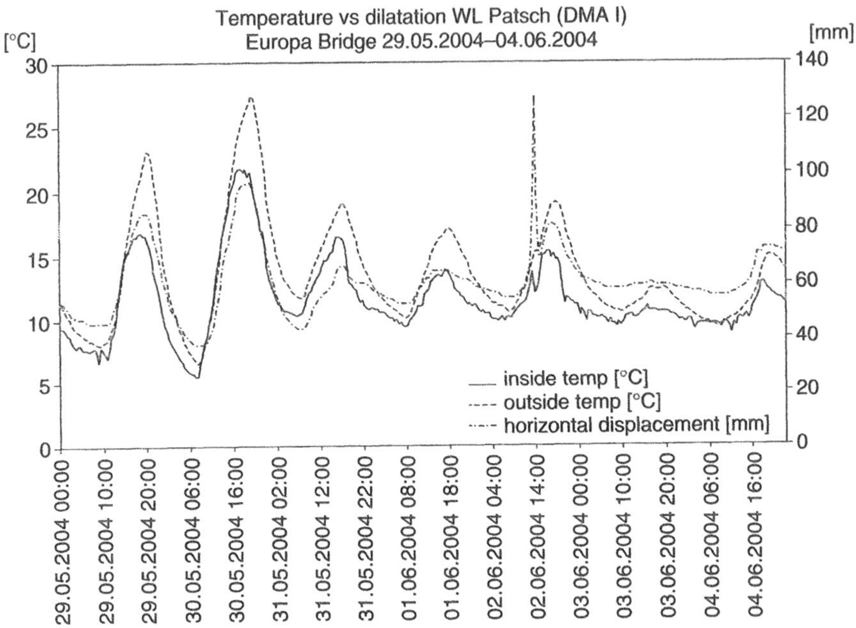

Figure 3.12 Focus on Figure 3.11 over a period of a certain week including one of the observed sudden reactions

reset forces was also recorded at a pre-stressed concrete bridge, namely the flyover at St Marx in Vienna (Figures 3.14 and 3.15).

The consequences of these records are that a realistic behaviour of a structure can be found through monitoring, which might explain damages in the outfitting. The displacements calculated for bearings and expansion joints might not be enough to cover extraordinary events as described. The centre of expansion of a structure can be dozens of metres away from the theoretical centre and influence the design of bearings and expansion joints (Figure 3.9).

3.2.5 LARGE BRIDGES VERSUS SMALL BRIDGES

In the beginning monitoring concentrated on large and important bridges. This has led to the impression that bridges normally perform very close to the theoretical behaviour determined and based on the design assumptions. The subsequent assessment of small bridges showed that it is considerably more difficult to achieve good results the smaller the structure is, because of different approaches taken towards these structures, which are not considered

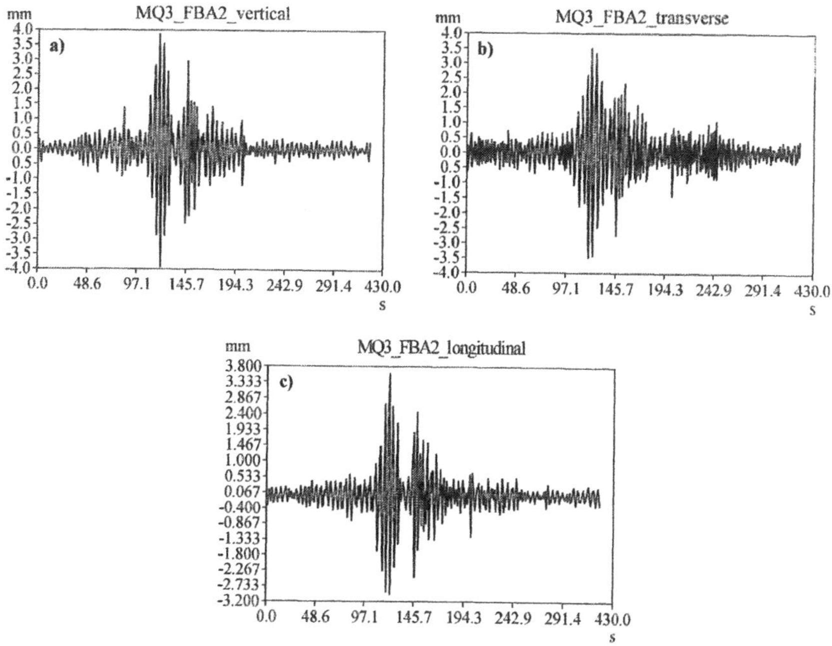

Figure 3.13 Relative displacement due to sudden occasions of restraint recorded with a three-dimensional acceleration transducer at the top of a 200 m high pier subdivided into the (a) vertical, (b) transverse and (c) longitudinal directions

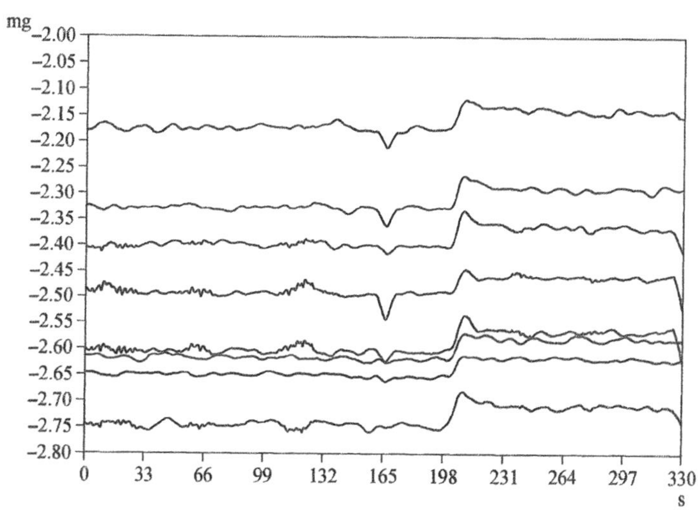

Figure 3.14 Displacement of the system's neutral axis due to bearing reset forces of the flyover at St Marx (basis: acceleration sensors)

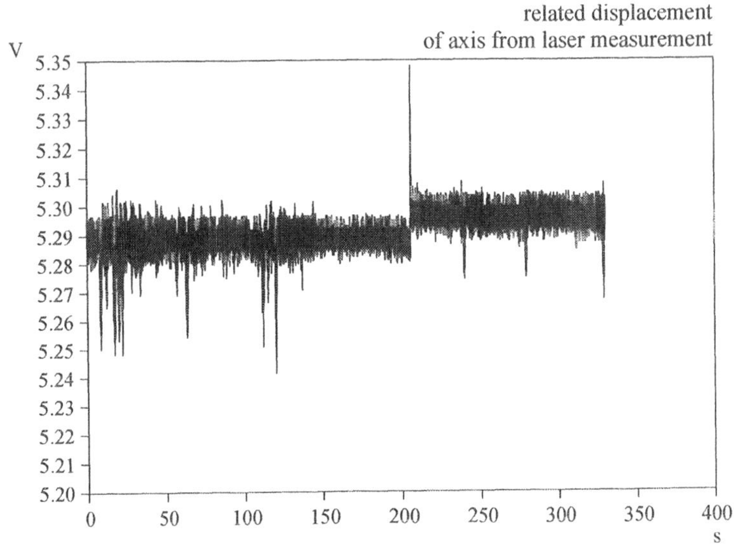

Figure 3.15 System displacement due to bearing reset forces of the flyover at St Marx (basis: longitudinal laser-displacement sensors)

as important. Another fact is that boundary conditions are much clearer in large structures.

The lesson learned from monitoring is that even higher attention should be paid to smaller bridges and that a number of provisions of construction codes fit very well for large structures but underestimate small ones. Here in particular the subject of temperature, as explained in another chapter, has to be highlighted. Furthermore, correct modelling of the boundary conditions has to be taken care of.

3.2.6 VIBRATION INTENSITIES

The subject of resonance in pedestrian bridges is well known and taken care of. Frequencies close to resonance, particularly those of structural members such as cantilever slabs, have not yet been properly considered. Experience has shown that an evaluation of the vibration intensities measured for a structure can give considerable information of fatigue and related problems. The assessment of vibration intensities can therefore give an indication of the expected lifetime of a structure and on local problems that might occur on structural elements in the near future. It has been clearly demonstrated that bridges where high vibration intensities have been recorded (Figure 3.16) most probably develop local problems in expansion joints, bearings, outfitting and, in particular, waterproofing. In Figure 3.17 the vibration intensity of a sound structure is represented.

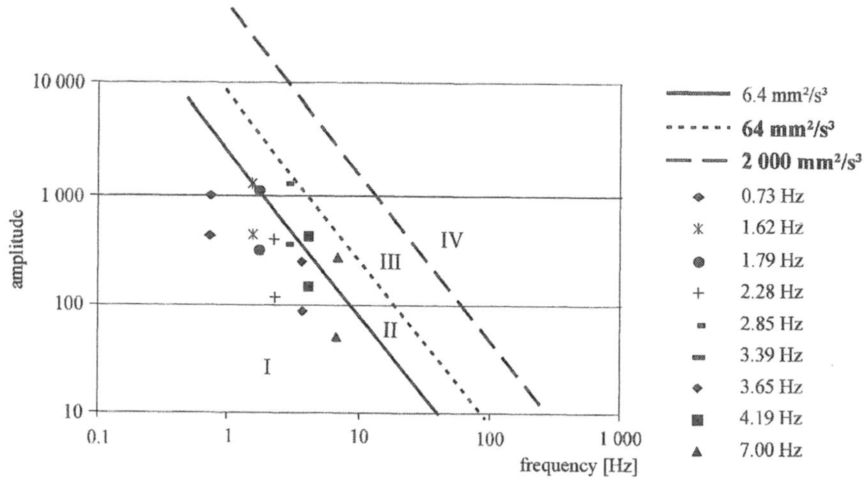

Figure 3.16 Intensity chart at the Europa Bridge of the Brenner Motorway (representing high vibration intensities)

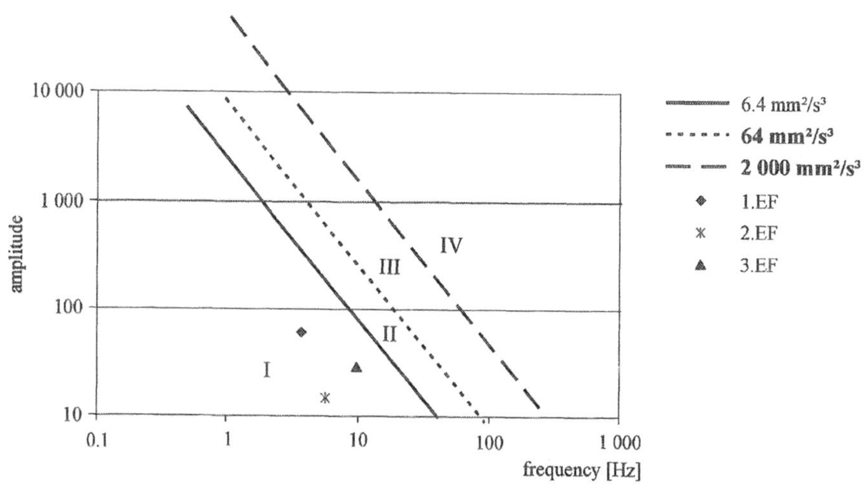

Figure 3.17 Intensity chart at the S 36 Bridge of the Al Motorway (representing low vibration intensities)

3.2.7 DAMPING VALUES OF NEW COMPOSITE BRIDGES

Measurements taken at a number of new composite bridges show that the damping values determined at a newly built structure are considerably higher than the normal values of comparable concrete bridges or steel bridges (Figure 3.18). This might be attributed to the fact that the composite effect becomes established through a number of load cycles. After some time the damping values of these bridges have been stabilized in normal ranges. Deeper conclusions on these phenomena have not yet been drawn, but it might be expected that a sharp drop in damping indicates eventual problems with bonding or the triggering of a hidden local damage.

3.2.8 VALUE OF PATTERNS

Certain elements of bridges exist in a repetitive form. All members are expected to show the same dynamic performance under service. One of the valuable

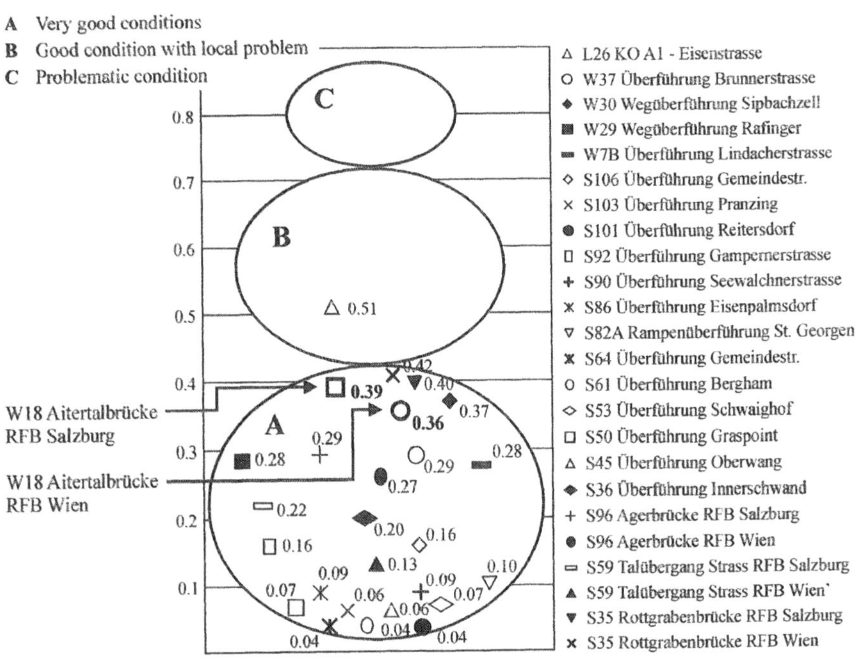

Figure 3.18 Classification of pre-stressed concrete and composite bridges according to their damping values

approaches of monitoring is to recognize patterns and to observe the performance of comparable components. Any deviation from the pattern indicates a malfunction or an extraordinary situation that can be identified.

As an example the case of a concrete box girder bridge with a distinct cantilever is shown. The monitored performance of the cantilever minus the action of the global system provides information on the cantilever eigenmodes. Related symmetric modes can be determined and displayed. This should provide a distinct pattern, where every deviation indicates a problem. On the basis of coloured frequency cards, so-called 'trend cards', the relevant cantilever eigen frequencies have been determined, which are marked in Figures 3.19 and 3.20. By comparing the response spectra of both box girders and their cantilevers (Figure 3.21), the share of cantilever vibration can be displayed directly (Figure 3.22).

A detailed evaluation procedure analysing the relation between the response spectrum and its energy content within the relevant frequency ranges leads to a certain behaviour pattern of the cantilevers along the bridge. Deviations from this pattern are typically indications of irregularity.

The following figures show the pattern of an undamaged cantilever compared (Figure 3.23) with a cantilever with minor corrosion damage of the transverse reinforcement (Figure 3.24). This method is not good enough for a detailed localization of the problem but it provides sufficient information on the quality of the function of a structural element. By a very quick and cheap test it can be determined whether action is required or not.

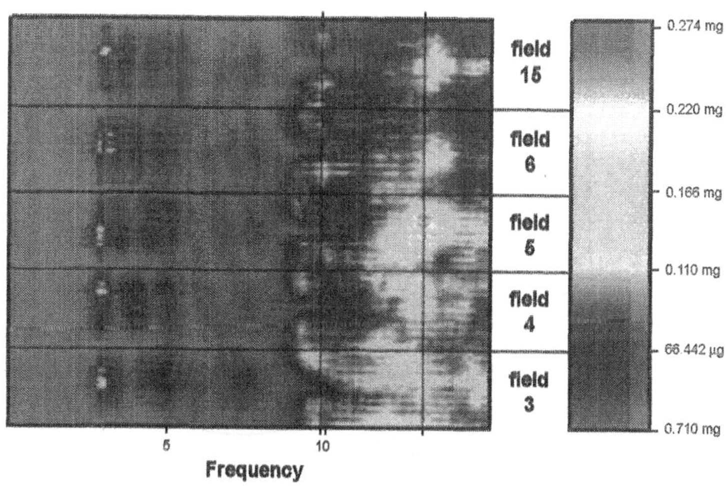

Figure 3.19 Course of frequencies at a certain concrete box girder bridge: structure south

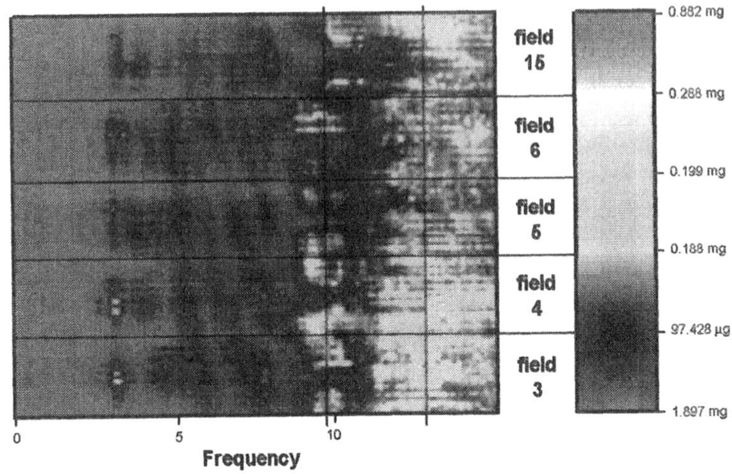

Figure 3.20 Course of frequencies at a certain concrete box girder bridge: structure north

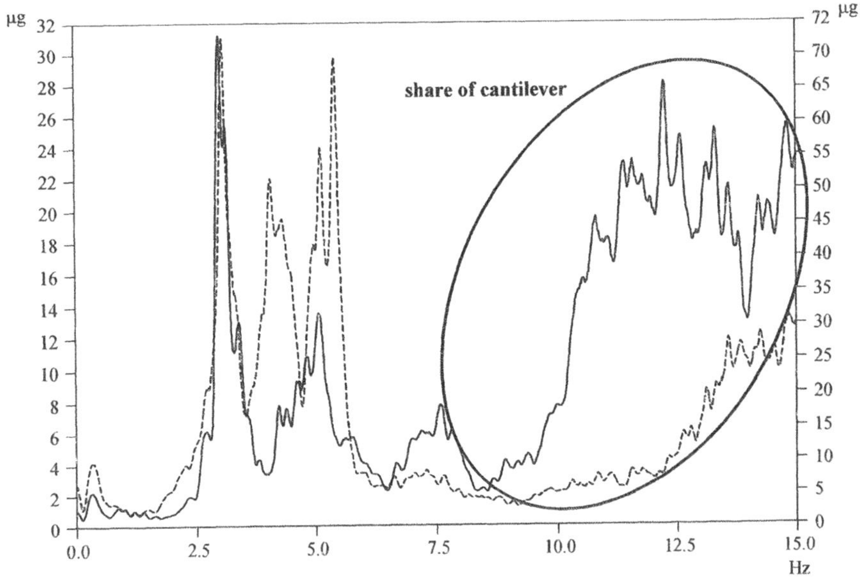

Figure 3.21 Spectrum of the cantilever (continuous line) and box girder (dashed line)

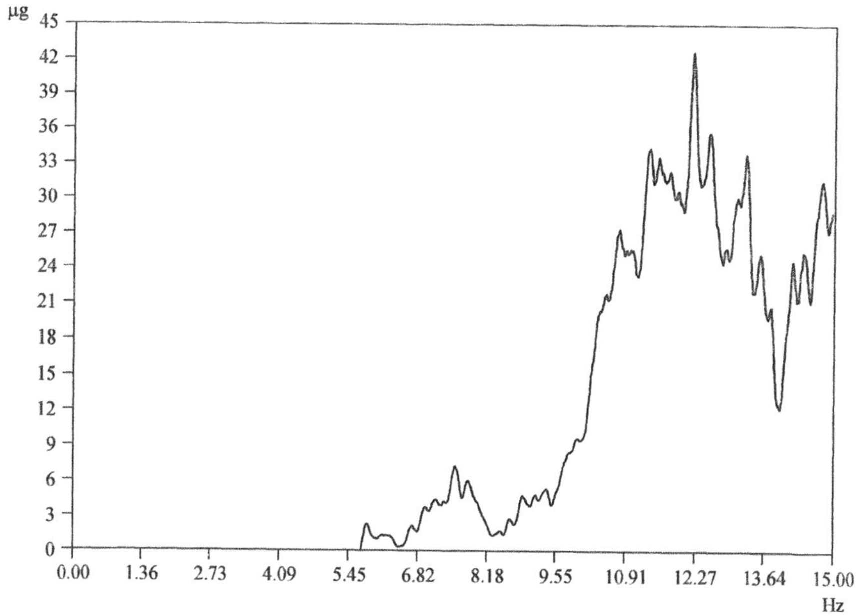

Figure 3.22 Response spectrum of cantilever vibration

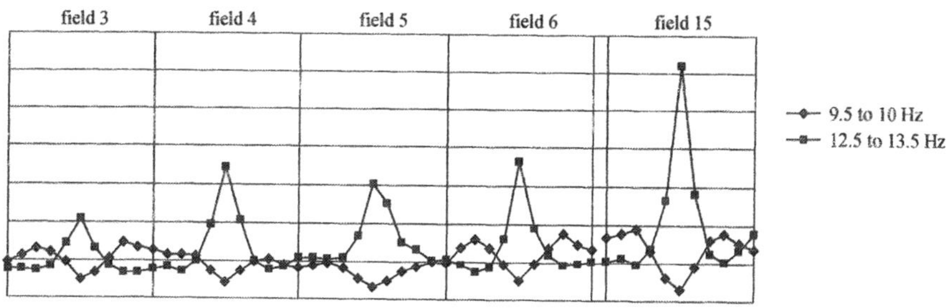

Figure 3.23 Acceptable behaviour pattern of the cantilevers along the bridge

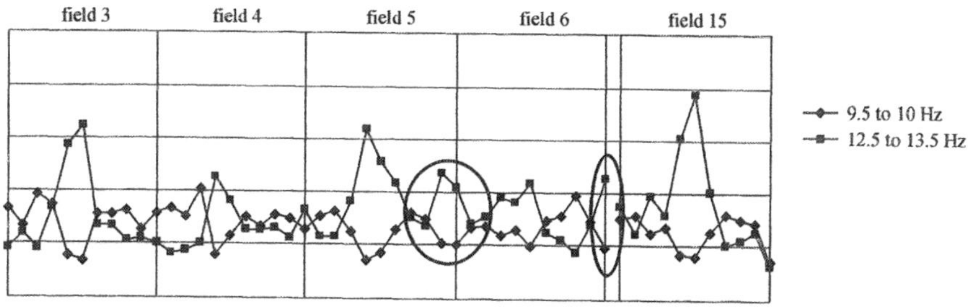

Figure 3.24 Behaviour pattern of the cantilevers with indications of irregularity

3.2.9 UNDERSTANDING OF BEHAVIOUR

Complex bridge structures are often modelled in a rather simple way, neglecting behaviour of the structure in the three-dimensional space. Monitoring is the recording of the actual behaviour of a structure. This comprises eventual drift or strain from temperature, as well as eventual construction mistakes, such as wrong placement of bearings or non-release of restrainers. The enclosed example shows a case where a temporary fixation during construction has not been removed at the time of the handover. The performance of the bridge has been considerably different from that estimated. Monitoring has been able to detect this difference and asked for immediate correction (Figure 3.25). Another important aspect is information on the actual displacement of a structure, particularly with regard to complicated cable-supported structures (see, for example, Figure 3.26), where such displacements could generate problems in traffic clearance or related interfaces.

Another way of finding problems is to compare the expected behaviour with the measured one. In the case of stay cables, protected by steel tubes against vandalism, the contact of the cable to the tube has been found through monitoring. The effective vibration length of the cable has been shortened by this contact. Such a problem can lead to drastic damage at a cable, providing a sharp edge that introduces unintended bending. Monitoring is able to identify these problems.

3.2.10 DYNAMIC FACTORS

Current bridge design codes ask for dynamic factors mainly depending on the bridge span. The factor is considered to be 1.40 for components or directly

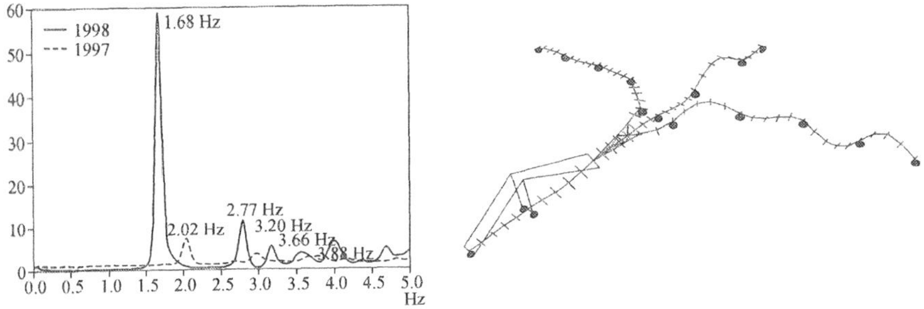

Figure 3.25 Frequency spectrum of Inn Bridge Hall West (1997–1998)

Figure 3.26 Steyregg Bridge

effected members and varies between 1.00 and 1.40 depending on the span of the bridge. The lessons learned from monitoring are:

- The dynamic factor provided by the code depending on the span length is actually conservative. All bridges so far showed smaller dynamic factors.
- The dynamic factor for components sometimes exceeds the values considerably. The record factor measured has been 2.20.
- The dynamic factor is very dependent on the speed of the vehicles. This can eventually be controlled by speed limits.

The consequences are that overloaded vehicles that drive on low speed will not produce harmful stresses. The low increase always has to be seen in conjunction with eventual speed effects. Consequently, dynamically sensitive elements should be avoided in the design.

Another lesson learned is that the dynamic behaviour also depends on the type of structure designed. Bridges with box girders (Figure 3.27) are considerably less vulnerable to dynamic effects than bridges with other types of design (Figure 3.28). The dynamic vulnerability of a structure depends on the acting mass. This is also clearly shown in monitoring records. Concrete bridges with a mass of $1.5\,t/m^2$ or more are not very affected by dynamic amplification. Continuous girders react less violently to any impact. Elements with major differences in stiffness produce an inharmonic behaviour that may damage the structure.

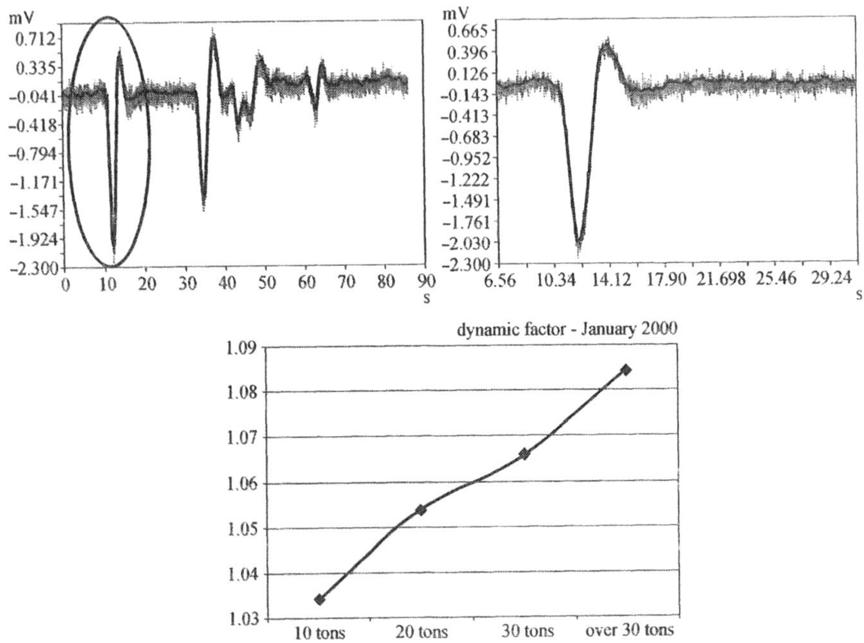

Figure 3.27 Dynamic factor of the flyover at St Marx

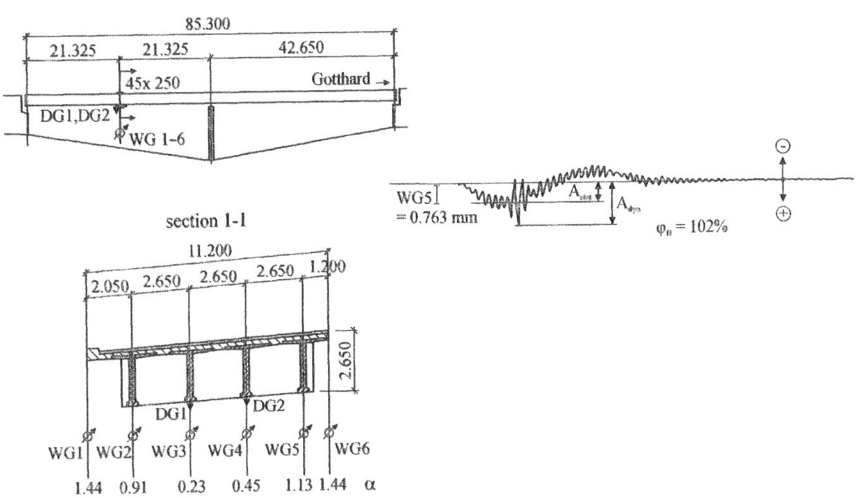

Figure 3.28 Dynamic factor of the Boeschrüti Viaduct due to induced impact loading (all dimensions in m) [3]

REFERENCES

1. Peeters, B. and De Roeck, G. (2000) *One Year Monitoring of the Z 24-Bridge: Environmental Influences versus Damage Events.* Proceedings of IMAC 18, The International Modal Analysis Conference, February 2000, San Antonio, Texas, pp. 1570–6.
2. De Roeck, G., Peeters, B. and Maeck, J. (2000) *Dynamic Monitoring of Civil Engineering Structures.* Computational Methods for Shell and Spatial Structures IASS-IACM 2000, Greece.
3. Cantieni, R. (1983) *Dynamic Load Tests on Highway Bridges in Switzerland – 60 Years Experience of EMPA.* Section Concrete Structures and Components, Report No. 211, Dübendorf, Switzerland.

4

Practical Measuring Methods

The assessment of the dynamics of structures has a long history. It has been known for a long time that conclusions on the condition of a structure can be drawn from its eigenfrequencies. As early as in the 1920s practical tests on steel masts were carried out in Switzerland. The practical utilization failed, however, because of inappropriate measuring technology. The chronology of the method is shown in the following list:

Nineteenth century	Development of the relevant sections of structure dynamics
1920–1945	Execution of simple tests at clearly defined structures
1955–1965	Further development of measuring technology
1965–1975	Development of the linear finite element method
1970–1980	Development of the forced vibration method, where the structure is excited by an external force to obtain vibrations
1975–1990	Integration of the linear finite element method into the general engineering design process
1980–1990	Promotion of computer technology
1990–2000	Integration of nonlinear finite element analysis into engineering practice
1992–1995	Introduction of the ambient method for vibration measurements
1993–1996	Introduction of computer measuring technology for data collection and, also for field measurements

Ambient Vibration Monitoring H. Wenzel and D. Pichler

Since 1994	Application of the ambient method by EMPA in Switzerland, by the province of Quebec in Canada and by EDI in Vancouver in Canada
Since 1995	Further development of the method by VCE
Since 1996	Commercial utilization by VCE
2000	More than 120 structures measured and assessed
2001	BRIMOS® recorder

4.1 EXECUTION OF MEASURING

Monitoring of technical systems is based on the knowledge of structural properties of the system to be observed. If system behaviour is sufficiently known, the sensors can be arranged at the right points of the structure. This enables an evaluation to be made of measuring quantities regarding relevant parameters for system characterization and condition assessment. The system findings obtained on the basis of theoretical modal analysis as well as the experience of the staff involved are the best prerequisites for the arrangement of a sensor plan for a structure.

Every load-bearing structure vibrates due to dynamic superimposed loads, while a 'quasi-stationary' structure reacts to excitations that are always present in nature. The minor vibrations of a structure due to these ambient excitations can be registered by modern highly sensitive acceleration sensors. The acceleration sensors must be arranged in such a way that a sufficient number of points along the system lines of the structure to be examined is covered for the determination of the modes. In particular, inconstant points (e.g. joints and coupling spots) need to be instrumented.

In most buildings it is necessary to examine more than nine check points, which can be simultaneously covered by currently existing acceleration sensors (Figure 4.1). For this the variable sensors are repeatedly rearranged, with a reference sensor always remaining at the same spot in order for the individual signals to refer to each other. The definition of the measuring grid and the selection of the check points is ideally done after a first calculation of the eigenfrequencies and the respective mode shapes in the computer model. Points on the structure, which reproduce the vibration shares of the individual mode shapes, can be instrumented, which allows the best possible identification of dynamic characteristics.

What has to be particularly considered for positioning the reference sensor is that a clear reference signal has also to be obtained for the identification of higher natural vibration forms (Figure 4.2). For this reason, for example, midpoints of the main field are unsuitable as reference locations because a node often already exists at the second vertical bending vibration point. Furthermore, it is an advantage if measurements are carried out at both sides of the

Figure 4.1 Acceleration sensors FBA-23 and FBA-11

Figure 4.2 Measurements taken on a six-span concrete bridge

structure in order to be able to identify torsion modes clearly. In most cases it is sufficient to have a sensor working in parallel in order to obtain information regarding the vibration behaviour at the corresponding eigenfrequency.

In order to guarantee efficient progress of the measuring series it is essential to study detailed documents regarding the structure in the run-up to taking measurements in order to be able to plan all respective steps. From the structural drawings a measurement layout can be established in which the sensor arrangement for each set-up can be unmistakably registered. In addition, the maximum cable lengths required need to be determined in order to find a good location for the basis station, on the one hand, and for the reference sensor, on the other. The position of the sensors should be recorded in a three-dimensional way, for the longitudinal development stationing should be registered and the distances to the structure axes should also be stated if possible.

For the specific questions in the scope of an assessment of a structure the characteristics of the object to be examined need to be considered when planning the measurement layout (Figure 4.3). It is particularly advisable to establish a denser sensor network at points like construction joints or links (Figure 4.4). The function of such load-bearing elements has to be clearly

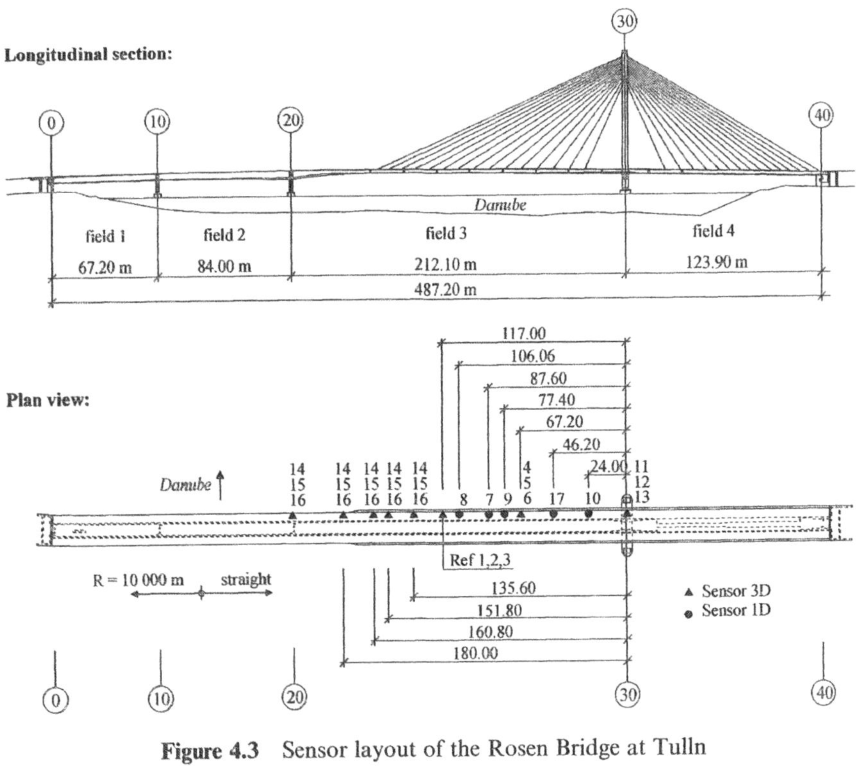

Figure 4.3 Sensor layout of the Rosen Bridge at Tulln

Figure 4.4 Measurement using a high density of sensors (local test)

examined for eigenfrequencies and the respective mode shapes. Critical condition modifications of the structure as, for example, sagging of supports or system changes can be identified and assessed using measurements of dynamic condition quantities. Support sagging, local and global stiffness changes as well as modifications in the static system always result in modified natural vibration behaviour. Resonant frequencies and the respective vibration forms are therefore observed by vibration measurements and their modifications are considered as indicators for structural change. Modifications of the vibration forms can also be easily identified in the ranges of nodes.

The measuring values required for implementation of an experimental vibration analysis are usually obtained from acceleration, velocity or distance measurements (laser) in the observed system. Accordingly, acceleration and vibration velocity sensors or laser systems are used for measurements. It is generally possible to calculate the variations in time of the measuring values from the variation in time of one measuring value by an integration or derivation and obtain the modal parameters from this. When selecting measuring sensors the following issues need to be considered:

- The eigenfrequency of the measuring devices has to be out of the measured frequency range in order to avoid undesirable secondary effects.
- In the case of ambient measurements on structures that do not show clear vibration behaviour, highly sensitive measuring sensors have to be used. These sensors will supply very accurate data, even in the case of very low eigenfrequencies, as they occur in large structures.

- The mass of the sensor must not be too large in relation to the contributing mass of the examined system because otherwise the dynamic behaviour of the system is influenced.
- The sensors have to fulfil their tasks reliably with regard to the rough operation conditions (humidity, temperature, mechanical stresses) and must show a corresponding long-term stability.

The sensors, in most cases acceleration sensors are used, have to be placed in the antinodes or the nodes if possible in order to be able to register their modification due to damage (e.g. constant support sagging). Furthermore, it should be remembered that the establishment and number of check points requires a visualization of the mode shapes from the measuring results. The following issues should be considered for the employment and the fixing of measuring sensors:

- When the sensors are linked to the system sufficient coherence with reality has to be observed. For frequencies recorded at a low frequency range, the dead weight of the sensors is mainly given. For measurements of vibrations taken at a higher frequency, fixing of the sensors can be required.
- The fixing of the sensors must not influence the stiffness of the structure to be examined.
- The eigenfrequency of the fixing construction must not lead to resonance phenomena in the observed frequency spectrum. The fixing construction to be carried out should be as stiff and, at the same time, as light as possible.

The basic rule for the scanning or sampling rate is that the minimum scanning rate should correspond to the fivefold maximum identified target frequency. This means that the scanning rate, e.g. for the highest noticeable eigenfrequency of 10 Hz, should be at least 50 Hz. This value has been verified for several measurements carried out until now and represents a clear increase in the required sampling rate compared to the well-known Shannon scanning theorem.

This criterion is necessary to be able to determine the frequency curve reliably. If additional information on short events is required, higher scanning rates have proved to be successful.

Finally, the scanning rate of 100 Hz has proved appropriate for registering individual events. The number of measuring processes was limited to 33 000 points (for FFT, $2^{15} = 32\,768$ points), which amounted to a measuring duration of 330 s per record. The files obtained a sufficient length to enable short-term disturbances of ambient vibrations to be cut out without loss of the quality of the results (Figure 4.5). The file size amounted to about 1.1 MB per measuring process when all 17 channels are used.

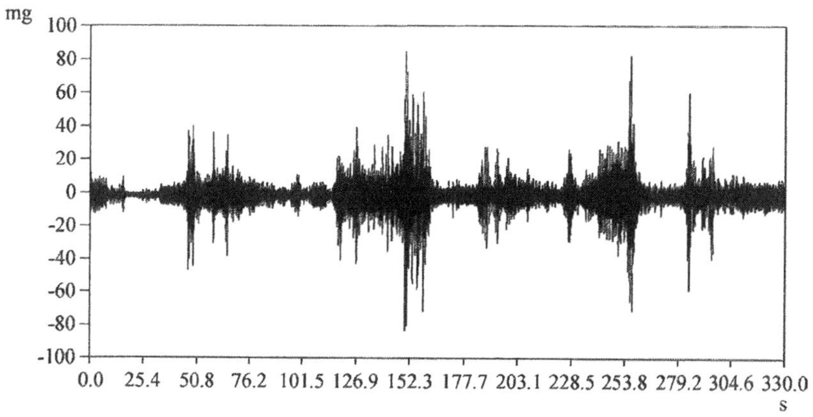

Figure 4.5 Vertical signal of all sensors applied (eight sensors)

Further important steps on the way to a fully integrated measuring system are:

- development of damage laws for concrete structures (time horizon of 4–5 years);
- further development of damage scenarios for steel structures;
- further development of the non-linear finite element calculation;
- further development of simulation techniques for the computer;
- further development of tests for prototypes for the computer;
- setting up of more accurate material laws corresponding to the requirements of dynamic methods.

4.1.1 TEST PLANNING

Detailed documents on the structure to be tested are required before the measurements are taken in order to plan all respective steps. From the plans a layout can be drawn up where sensor installation can be recorded. Furthermore, the maximum required cable length is to be determined and a convenient location is to be found for the reference sensor. The position of the sensors is to be registered three-dimensionally. For longitudinal development, stationing is to be registered as well as the distances to the axes.

4.1.2 LEVELLING OF THE SENSORS

The sensors have to be levelled each time before being put into operation and at periodic intervals. This is done be setting all nine sensors up at the same location and comparing the signals with each other. Then an immediate correcting measure can be taken in the case of disturbances.

4.1.3 MEASURING THE STRUCTURE

The structures need to be surveyed in such a way that the setting up of all sensors is possible at the correct projected points. In the case of a curved structure the change in length of both sides has to be considered as the sensors are mainly established at the margin, not at the axis. The location of the reference sensor has to be marked as clearly as possible so that it can be easily found for measurements that last for several days.

4.2 DYNAMIC ANALYSIS

Computer models for the simulation of the behaviour of real structures have to represent a realistic image of the actual load-bearing capacity in order to provide an accurate result. Numerous computer applications allowing the establishment of FE models as well as the adaptation of the model to the measured results are currently available. Such calculation models can be used to examine the elastic behaviour of a structure under various load conditions (static or dynamic). Furthermore, it is possible to simulate the effect of reconstruction or rehabilitation concepts under realistic conditions using the calculation model. This approach leads to a far better understanding of the function and load-bearing conditions of the system.

4.2.1 CALCULATION MODELS

In most cases a relatively simple model is sufficient for modelling a structure. For this reason in most cases a framework program (I.c. Rstab by the Ing.-Software Dlubal GmbH) is used for determination of the dynamic structural parameters by calculation. Such a program has efficient dynamic modules, which enables the calculation of natural vibrations as well as of forced vibrations. In addition there is the possibility of using a program that has provisions for modelling shell structures or plates, since such systems cannot be realistically described by a framework program. A prerequisite for a mathematical system analysis is an adequate calculation model. In particular, structure stiffness and masses as well as the structure bearing have to be registered sufficiently accurately in order to be able to calculate the modal parameters realistically. The damping parameters cannot be determined in the course of such a calculation. In this case a comparison with experience and literature values as well as a quantitative comparison of individual values on the longitudinal development of a structure need to be used.

As calculation models, in particular of damaged structures, are very complicated an analytical solution method is not applicable; therefore numerical solution methods are predominantly used in practical operation. In most cases the finite element method (FEM) is used for this purpose. Examples of models are shown in Figures 4.6, 4.7 and 4.8. When the calculation model is adapted to the results of dynamic measurement, the modelled building must be modified. In this process local phenomena such as cracks or damage to the footings and mountings have to be taken into consideration, as well as global problem areas (crack areas, modification of the E-Module).

Problems with the modelling of structures could be inaccuracies that occur during the definition of the structure. This refers, for example, to the selection of the element network and the type of elements as well as the boundary conditions of the structure. Materials, flexural and spring stiffnesses as well as masses and the moments of inertia of the individual elements are particularly important for the selection of individual parameters. The aim of modelling the structure is to lose as little information as possible despite far-reaching simplification of the complex building for the input of the structure. This is decisive

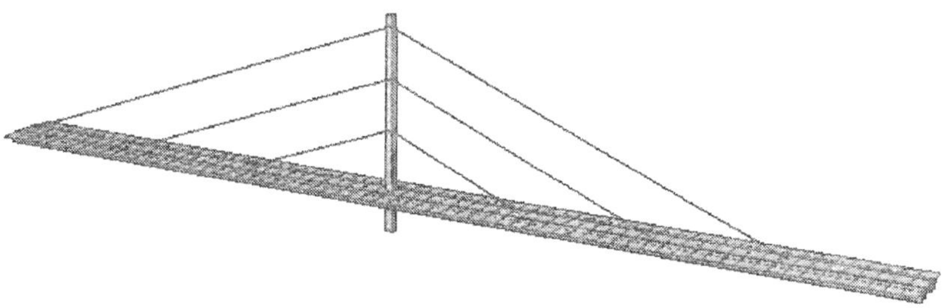

Figure 4.6 Beam model of the Voest Bridge at Linz

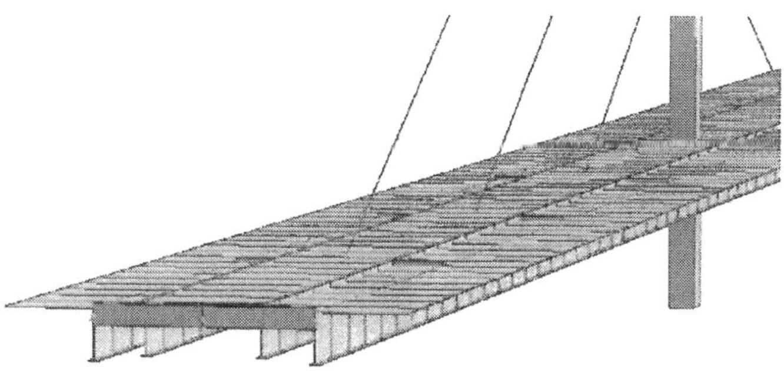

Figure 4.7 Details of the cross-section of the Voest Bridge at Linz

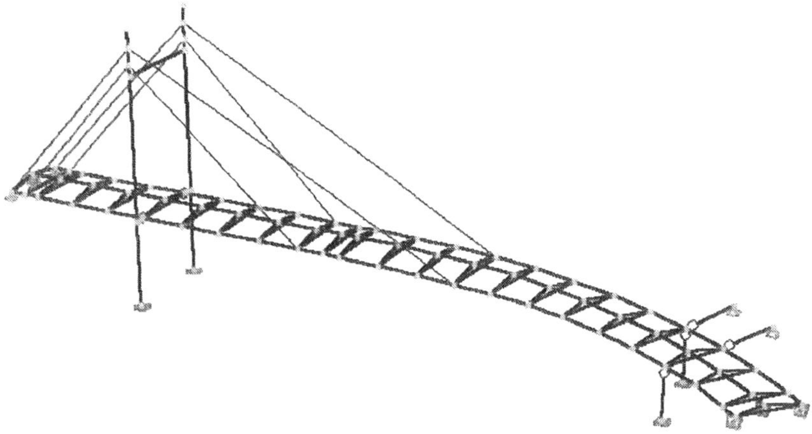

Figure 4.8 Beam model of the Lech Bridge at Vils (composite system)

for a realistic calculation of dynamic parameters like eigenfrequencies and mode shapes. Mass distribution of the structure is a decisive point when deciding the magnitude of the calculated eigenfrequencies.

Furthermore, the method is required to draw up correspondingly adapted calculation models for different load-bearing systems. Whereas it is, for example, possible to simulate a box girder with a single bar, this simplification does not supply satisfactory results for a double T-beam. The dynamic load-bearing behaviour is completely different in reality. If this basic prerequisite is not taken into consideration, no good results can be expected during the determination and comparison of the eigenfrequencies and mode shapes. For a double T-beam it is necessary to show every T-beam as a single bar with its contributing slab width. Consequently, a bending-resistant connection of the two T-beams has to be defined. This issue is significant for the modelling of composite systems as a shear-resistant connection between the steel girder and concrete slab that is too weakly dimensioned will lead to wrong results (missing system stiffness).

This approach for replacing a structure with a single bar has to take into consideration the following points. As torsion forms have to be determined by mathematical analysis, loads must not be attached exclusively in the central axis of the system but need to be distributed over the structure cross-section corresponding to the real structure. This requires the insertion of crossbars, which are connected as cantilevers starting from the central bar. They are joined with a very high flexural stiffness in order to avoid natural vibrations. At the ends of these additional bars a part of the structure mass is applied, with the remaining difference to the real structure mass arranged at the central bar. Realistic results can therefore be obtained by means of a simple model. This approach shows that the defined crossbars must be included into the mass determination, which should be established from weightless media. In the following example, namely the Danube Bridge at Hainburg (Figure 4.9), the steel box is depicted by numerous longitudinal bars, to which the corresponding statically effective cross-section values are assigned. The masses of the cross-section including the bridge extension as well as the individual stiffeners are considered in the mass distribution of the structure (Figure 4.10).

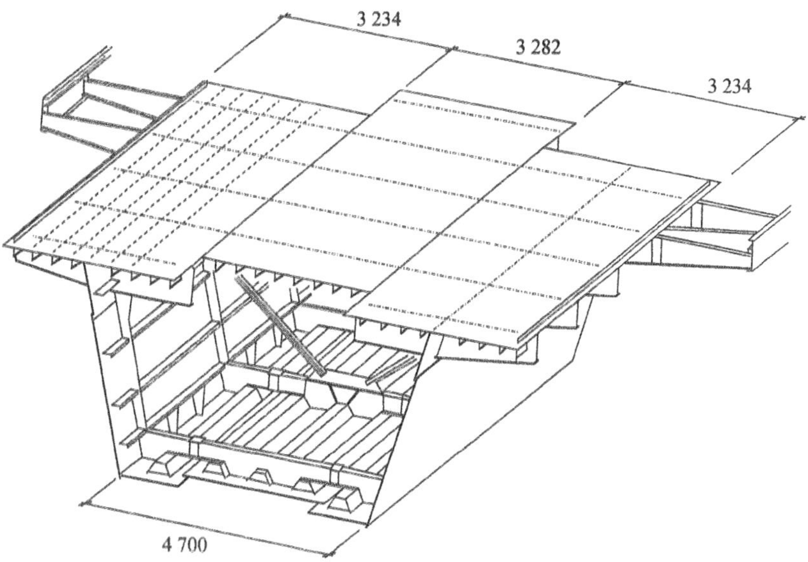

Figure 4.9 Cross-section of the Danube Bridge at Hainburg (all dimensions in m)

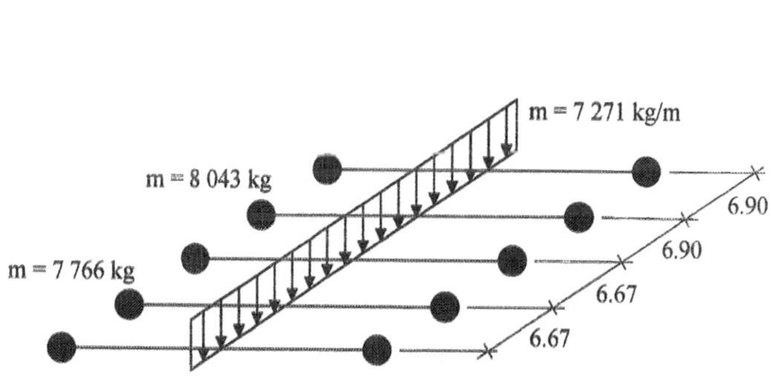

Figure 4.10 Model of the bridge (all dimensions in m)

4.2.2 STATE OF THE ART

The development of computer technology can be seen from Figure 4.11. Process velocity has increased from 0.01 mega flops (million floating point operations per second) in 1955 (IBM 704) to 1 000 mega flops today. Current computers are approximately 100 000 times faster than the devices used in 1955.

A dynamic analysis is needed to draw up a realistic model of the structure. By means of modern FE programs it is possible to model structures in almost unrestricted accuracy, limited only by economical restraints. Experience has shown that it is more important to place masses correctly than to design networks as large as possible. Therefore programs have to be found that operate directly with cross-sections from designs, allow a representation of the structure and make a check of the possible inputs. For an assessment of the global condition of a structure it is sufficient to arrange ten elements per field. For local examinations more refined models have to be applied, also using slab and shell elements. For a representation of extreme mechanisms non-linear programs can be used as well, but an improvement of the global result is not expected. Calculations that depict very detailed local structural parts are very extensive and thus can only be supported in a few cases owing to their economic cost, but they might be interesting for scientific reasons.

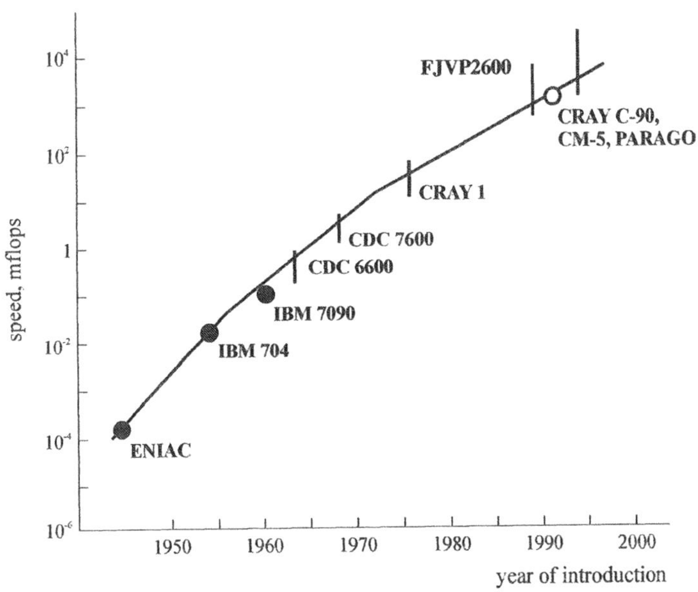

Figure 4.11 Development of computer velocity

The costs of computers have decreased by approximately four thousand times between 1966 and 1996 for the same calculation performance. For non-linear computer analysis the following facts should be taken into consideration:

- Systems with up to 100 000 000 degrees of freedom have already been calculated.
- In the case of non-linear calculations there is always a particularly high dispersion. This makes interpretation difficult. It is always important to understand the processes with regard to the occurring phenomenon before carrying out the calculations. Therefore the AVM should be applied by practical engineers.
- An approximation of $\pm 10\%$ to reality is regarded as an excellent result.
- In most cases it is not sufficient to carry out single calculations, but the problem should be checked with various systems under different assumptions and marginal conditions. The different models subsequently have to be interpreted.
- Realistic material and damage detection laws will exist in three to four years at the earliest. Widespread employment in practice will probably take five more years.
- In many cases it is considerably easier to solve problems dynamically than by a quasi-statical approach because the equations are better defined for the dynamic approach.
- The current solution algorithms for matrixes work with the elimination of factors that could possibly play a role for the eigenfrequencies. This phenomenon has to be checked in future.
- In reality most problems are non-linear. What differs is only the degree of non-linearity and the importance of non-linear considerations.
- In the case of non-linear calculations it is particularly problematic to 'blindly' trust in the computer and the program (i.e. without verification).

4.3 MEASURING SYSTEM

The AVM is a closed subjective methodology depending on the systems used. Success is related to certain structures of measurement. Therefore a successful application will be described here.

4.3.1 BRIMOS®

BRIMOS® (bridge monitoring system) has developed from long-standing monitoring activities of VCE (Vienna Consulting Engineers). The most important development steps are chronologically listed in the following:

BRIMOS® 1.0	Tension measurements in tunnel shells and concrete structures (underground railway construction at Vienna, Olympic Grand Bridge in Korea, 1989)
BRIMOS® 2.0	Static monitoring of cable forces at the cable-stayed bridge in Tulln, monitoring of cracks as a result of underground railway construction (1993)
BRIMOS® 3.0	Development of frequency analysis based on FAMOS (fast analysis and monitoring of signals), FFF (Forschungsförderungsfond) project VCM (vibration characteristic method), switch to a multi-channel system, development of ANPSDs (1996), EMPA (Eidgenössische Material prüfungs -und Forschungsanstalt) training
BRIMOS® 4.0	Introduction of RDT (random decrement technique) for the determination of damping, calculation of mode shapes, automatic evaluation of data records (1998)
BRIMOS® 5.0	Laser calibration, generalization of input data (channel allocation), animation of results, MAC (modal assurance criteria) assessment, trend analyses, intensity analysis (1999)
BRIMOS® 6.0	Change to programming language C+, development of the BRIMOS® recorder (2001), development of classification according to BRIMOS®
BRIMOS® 7.0	Automatic data import from measurement layout, automatic sensor calibration before every measurement, determination and animation of the mode shapes, trend investigations of the eigenfrequencies, improvement of RDT, graphical result illustration, system was equipped with new sensors (2004)

The BRIMOS® system consists of measuring the vibration behaviour of the examined structure, on the one hand, and of an analytic part – the comparative calculation with the computer model – on the other hand. As already explained, dynamic characteristics are recorded by acceleration measurements at the structure. Figure 4.12 gives an overall view of the measuring equipment used. The conception has been based on works carried out at EMPA in Switzerland.

4.3.2 SENSORS

The sensors FBA-23 and FBA-11 – a product of the company Kinemetrics (USA) – measure accelerations with a sensitivity of 10^{-6}–1 g. The measuring adjustment for ambient structure vibrations is done in such a way that 1 g corresponds to 2.5 V. The principle of measuring can be compared to an electronic spirit level where a mass is held in an electric field and its deviation from the zero position is measured. The maximum resolution amounts to

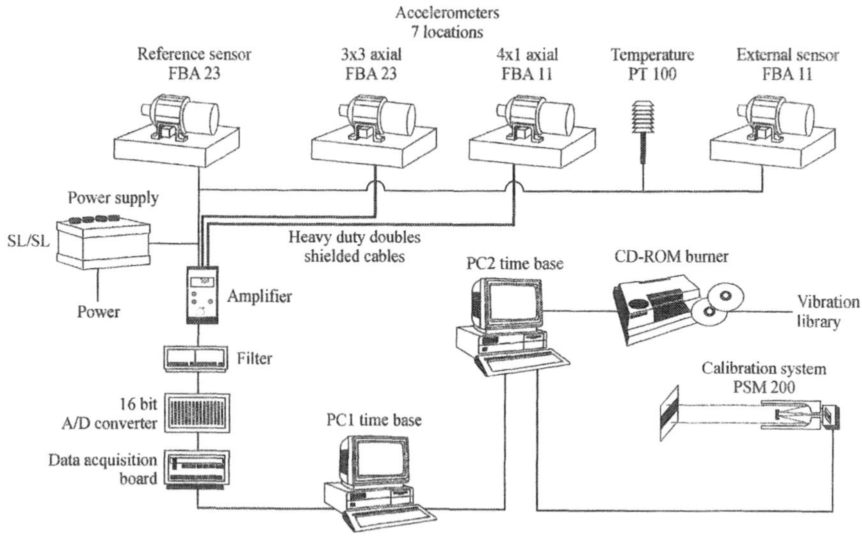

Figure 4.12 BRIMOS® system configuration

10^{-6} g and is therefore very high. The sensors are mounted on aluminium cubes in order to avoid natural vibrations and to enable accurate levelling. The cubes were constructed in such a way that their dead weights prevent a displacement of the sensors under dynamic stress. They are installed by means of adjusting screws which can be fixed with counter nuts. Every sensor is marked with numbers to enable allocation of the levelling signal.

4.3.3 DATA-LOGGER

The analogous data measured by the sensors are passed on to the data-logger μMusycs of IMC Germany via specially protected cables. The transfer and saving of raw signals is carried out by means of a 16 bit AD (analogue to digital) converter. Special cables with 12 strands and double shielding are used, which cause very low losses during transmission and are excellently shielded against external magnetic fields.

In order to eliminate interfering signals from electric fields, for every metre flume an analogous low-pass filter is superposed on the data-logger. It can be regulated in stages from 50 to 0 Hz. These filters can eliminate forced vibrations in the high band range.

Signal amplifiers are used to amplify very small ambient signals, thus enabling a reliable evaluation. The amplifiers of the model series AMP-11 by Kinemetrics provide a double amplification. This gives a very sensitive amplification. In the

Figure 4.13 Measuring facility in the vehicle

basic set-up amplifiers are not used; they are employed for especially stiff and massive structures only.

The measuring data are recorded by an industrial PC (in order to avoid pollution), which is equipped with a Pentium processor and a big hard disk as large amounts of measuring data have to be processed. To enable an efficient way of data transmission, the system is equipped with a network connection enabling data transfer to a notebook. Furthermore, it is possible to carry out on-site FFT short analyses by means of this parallel connection in order to examine the suitability of the measuring data. The entire measurement equipment can easily be stored in a van, as shown in Figure 4.13.

4.3.4 ADDITIONAL MEASURING DEVICES AND METHODS

If in addition to the frequencies the actual vibration amplitudes are required, a laser placed outside the system is applied and remains stationary, since by laser measurement more accurate data are obtained instead of displacement determination by means of operating double integration on acceleration signals. The laser beam is aimed at a reflector on the structure, which is connected to the BRIMOS® system. The movement of the laser point is measured digitally. The results are deformations with an accuracy of $\frac{1}{100}$ mm.

For complex tasks it is often necessary to carry out additional measurements. In particular, for mass–spring systems in railway tunnels, displacements with very small amplitudes are interesting. Therefore highly sensitive displacement sensors are installed that only measure static deformations. At the same time it is necessary to determine the respective structure temperature in order to register changes in lengths that are dependent on the temperature.

It has proved successful to use vibration velocity sensors for the assessment of sensible vibrations instead of acceleration sensors. They yield very good results for determination of the K_B value as well, but do not have the high dynamic measuring range (≥ 0 Hz) required for health monitoring. Comparative measurements have shown that the vibration characteristic is only represented sufficiently accurately by the force balanced accelerometer (FBA) of Kinemetrics.

For vibrations with higher intensities so-called 'piezoelectric' sensors can be used. They are considerably cheaper but have large deficits in the important low-frequency range (0–5 Hz). They can therefore not be reasonably applied for ambient system identification. They are, however, appropriate for applications with cables and vibrations of structural members at higher frequencies.

4.4 ENVIRONMENTAL INFLUENCE

Environmental influences such as temperature and humidity mean considerable stress for a structure as heavy additional loads can be introduced into the structure. The latter have an influence on other values, such as expansions, crack behaviour as well as dynamic properties of the structure. Furthermore, environmental influences can result in a direct damage potential (corrosion) via chemical influences.

If measurement of the dynamic behaviour is called for in an assessment of the condition of the structure and if so-called 'health monitoring' is aimed at, it is necessary to distinguish between normal changes of the dynamic behaviour and extraordinary changes (damage). Normal changes in the dynamic behaviour of a structure are caused by variations in environmental conditions like humidity, wind and, as a decisive factor, temperature. The temperature exerts a decisive influence on the boundary conditions of the structure as, for example, frozen ground or the E-Module of the building material. Extraordinary changes of the dynamic behaviour are caused by a loss of stiffness through damages (e.g. crack formation) or by changed bearing conditions.

It is obvious that normal variations in dynamic behaviour, e.g. caused by the climate, should not lead to a false interpretation of the actual condition of the structure and the degree of damage. Normal variations exert a harmless influence on the load-bearing behaviour, whereas extraordinary changes can lead to a critical condition for safety.

In the widest sense exterior influences, e.g. traffic loads acting on a structure, are to be counted as environmental influences. Then the induced loads from the dynamic reaction of the structure have to be recorded and their influence on the modal parameters assessed. The following tasks can be determined by the dynamic behaviour of the structure:

- Continuous observation of the loads from traffic by measurements. These evaluations can also be used for verification of existing load models and the drawing up of alternative load models.
- Determination of load collectives and vibration coefficients by acquiring the acting stresses according to type, location, size, duration and frequency. Influences from wind and temperature can be additionally considered. Such load models allow more precise statements to be made on the operation stability and the remaining service life of endangered structures or structural members.
- Improvement of load models that can only be estimated with difficulty in the design phase.
- Conclusions on influences of the surroundings as aerodynamic vibration excitations and temperature progress.
- Derivation of specific measures for stress reduction by modification of the stress type or of construction.
- Statistics on the long-term trend regarding the increase and decrease of traffic loads.

The environmental conditions, especially the effect of temperature and traffic on the structure, influence the results. Therefore an accurate knowledge of these effects is required. The influence of temperature has already been proved by continuous measurements taken at several structures. The results show that the change in frequencies due to temperature changes essentially depends on the total stiffness of the structure. For stiff flyover structures (e.g. Z24 in Switzerland from the SIMCES project) the change amounts to approximately 1% per 10 °C temperature change. For more flexible structures, like the Olympic Grand Bridge in Korea (cable-stayed bridge), this influence only amounts to 0.2% per 10 °C temperature change (Figure 4.14). What has to be particularly considered is, however, the phenomenon of stiffening in the case of negative temperatures. An extra study series is to be dedicated to this phenomenon and its effects. In the BRIMOS® program temperature compensation was installed where the temperature of the structure is registered at the time of measuring and the corresponding value at a temperature of +15 °C is calculated. The structure stiffness has to be assessed by the user, but possible mistakes are, however, acceptable.

When temperatures were measured at various structures useful influence curves could be determined. Figure 4.15 shows the temperature curve of a mass–spring system that is not directly sunlit and therefore experiences constant changes over the year. From the figure it can additionally be seen that the annual curves are very similar for the observation period. Here the approach for temperature compensation is to be found.

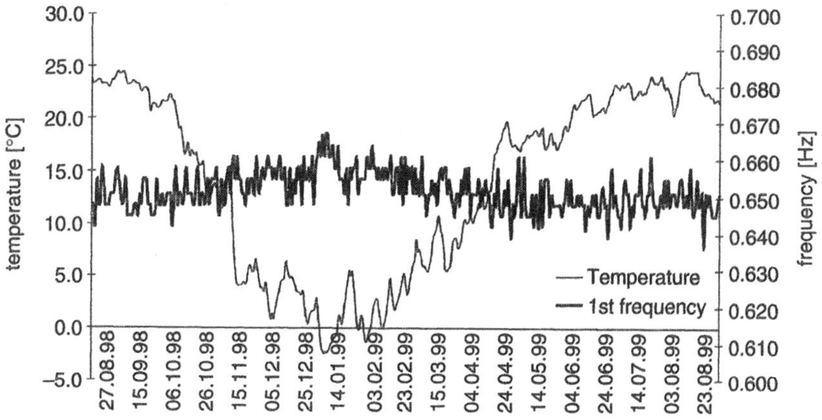

Figure 4.14 Temperature versus stiffness curves at Olympic Grand Bridge (one year)

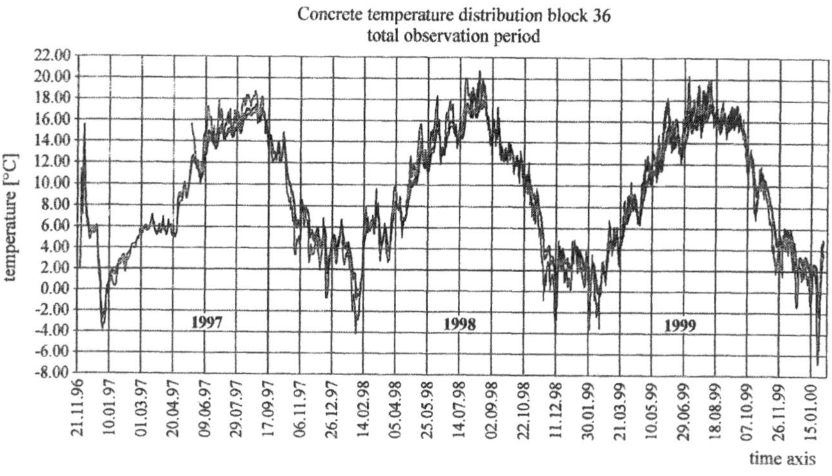

Figure 4.15 Temperature curve of Römerberg Tunnel (three years)

The influence of additional loads from traffic can usually be neglected. Due to the measurement method only very heavy and in particular static loads are significant. These special cases can, however, be considered separately by also applying the load in the mathematical model. The proof of this influence was found from a 24-hour test on Nord Bridge in Vienna, where the relevant frequencies only imperceptable changed despite the passage of 76 000 vehicles (Figure 4.16). It is therefore permissible to assume that the stress was white noise. Individual events can be filtered out by a selection programme, which determines special events that were not noticed during the measurement.

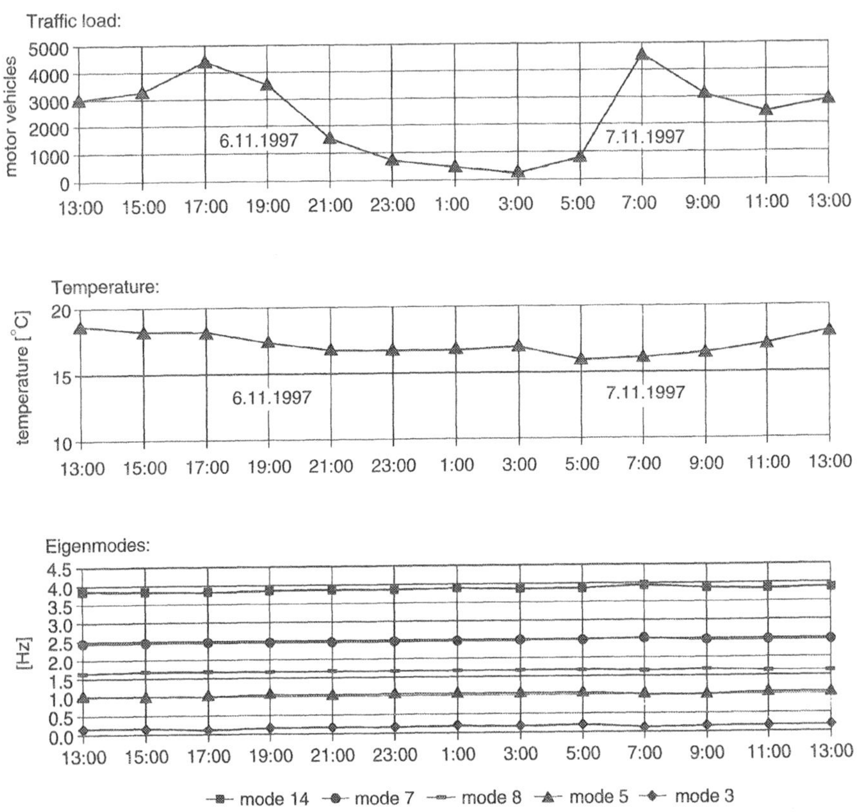

Figure 4.16 Relationship of traffic load, temperature and vertical natural modes

4.5 CALIBRATION AND RELIABILITY

The sensors of the measuring system supply a correct image of its characteristics, which, however, show only relative values (for the relationship between each other) for deformations. In order to record displacements and obtain exact deformation values, the signal has to be calibrated.

The laser beam is aimed at a reflector mounted on the structure and yields measurements of the occurring deformations in millimetres. A calibration of the displacement signals determined by double integration from the accelerations can be performed. During the set-up of the system the laser (Figure 4.17) must be installed at a steady point away from the structure so that no other deformations distort the data. The laser measurement must be carried out simultaneously with the acceleration measurement so that identical events can be evaluated.

Figure 4.17 Laser unit

It would be advisable to establish accordance with the reference sensor. For this reason the cube of the reference sensor is equipped with a corresponding panel to which the reflector can be fastened.

4.6 REMAINING OPERATIONAL LIFETIME

Bridges are ageing and traffic is growing, which creates a demand for accurate fatigue life assessment. This section deals explicitly with steel bridges and shows how to utilize today's monitoring abilities, which enable performance to be measured precisely. The procedure described could be used in the following situation.

A representative steel bridge is to be observed as the assessed prevailing vibration intensities might cause fatigue problems and possible damage. The combination of measuring and analytical calculation on the assessed bridge leads to a detailed system identification and is of crucial importance for the layout of a permanent measuring system.

The superior goal is to determine the relation between the randomly induced traffic loading (vehicles per day) and the fatigue-relevant, dynamic response of the structure. As lifetime predictions in modern standards depend on many assumptions, the emphasis is to replace many of these premises by measurements. Most of today's large motorway steel bridges have superstructures represented by a steel box girder and an orthotropic deck and bottom plate.

In long-span bridges the load on the primary superstructure is dominated by the dead load. As the fluctuating live load part is relatively small, fatigue is of secondary importance. The deck, stringers and floor beams are mainly subjected to live load and therefore they may be controlled by fatigue. Consequently, the focus will be on three levels:

- global behaviour dependence on all relevant loading cases;
- cross-sectional behaviour under special consideration of the cantilever regions;
- local systems analysing the interaction between tyres and the beam-slab connections.

In each of these levels of analysis consumption of the structure's overall capacity per year is to be determined.

The ability to merge high-precision sensor data of acceleration and displacement dependence of separately recorded wind and temperature data provides the possibility to realize lifetime considerations, which are of eminent importance for bridge operators and users.

The research work discussed in the present chapter started with the installation of a permanent measuring system on a famous Austrian highway steel bridge during September 2003. As the relevant authorities are not intending to publish quantitative statements at the present state of investigation, the methodology in general with all its different sources of input and further future work are discussed in the following.

4.6.1 RAINFLOW ALGORITHM

The elapsed time of the structure's response owing to a randomly induced traffic load is recorded by high-precision sensor data. An indispensable requirement is to reduce the enormous amount of information of the permanent measuring system to a few statistical data for further assessment. The rainflow counting method reduces the sensor data's complete load–time history, represented by random sequences of peaks and valleys, to a set of fatigue-relevant recurring response cycles in different categories of intensity and occurrence, and has become state of the art in fatigue analysis related to non-periodic loading. The analysed random time series may be considered as matching pairs of reversals: the reversal from the maximum to the minimum in the signal and the reversal from the minimum to the maximum, with all the other reversals effectively interrupting these two. This phenomenon is often called the 'material memory', because a material subjected to a sequence of reversing loads apparently interprets each closed cycle (matching pair of reversals) as a temporary interruption of a larger strain range, and remembers which complementary hysteresis part applies for this larger event.

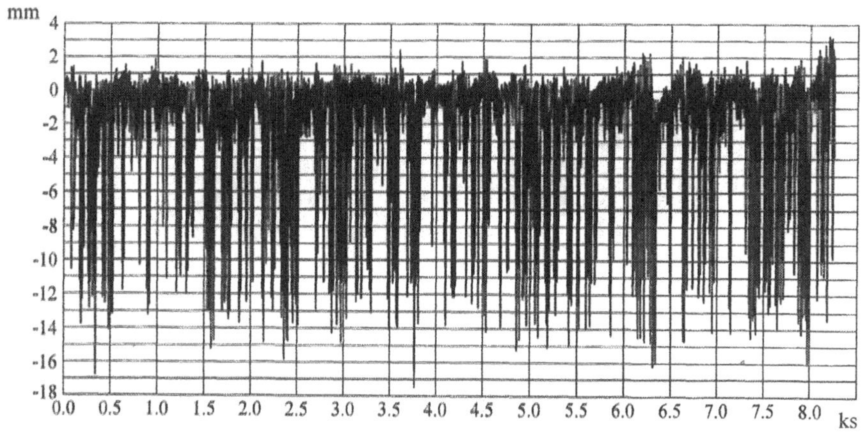

Figure 4.18 Cantilever's response owing to predominantly occurring truck traffic

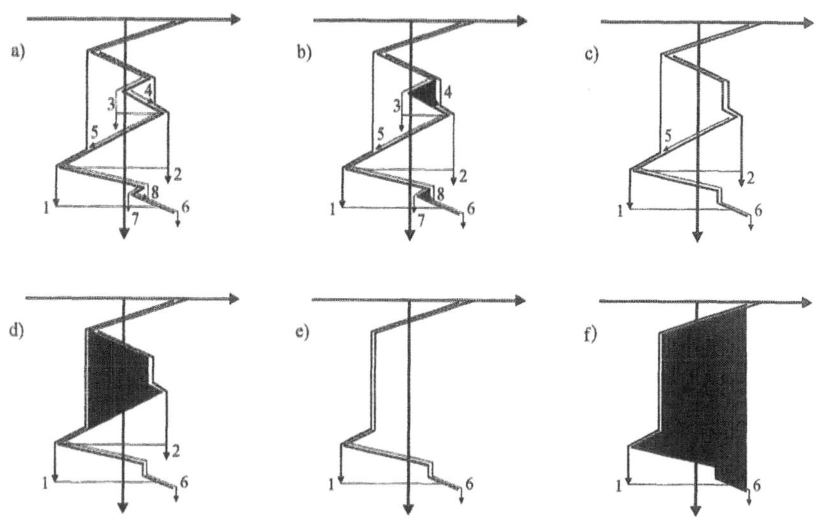

Figure 4.19 Rainflow counting example

The algorithm used by the author is based on reference [1]. The displacement history (Figure 4.18) is rotated by 90° in the clockwise direction. An imaginary flow of water is then initiated in every peak and valley. For a better understanding focus is made on the short sequence given in Figure 4.19. A certain flow (specified as number 2) is then followed until it experiences a drop:

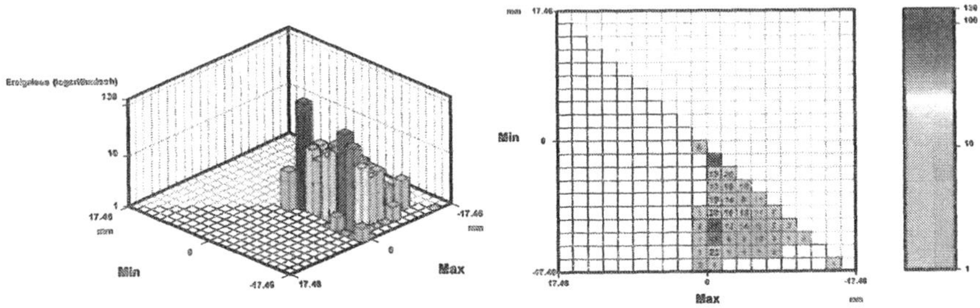

Figure 4.20 Axonometric projection of the rainflow matrix (left) and its relating ground view (right)

- If the observed flow intersects a second flow originating from a peak or valley of a smaller absolute value than the origin of the first flow, then a cycle can be counted. The cycle ranges between the values of the drop-off point and the origin of the second flow (Figure 4.19(b): closed loops 3 and 4; 7 and 8 are identified and saved).
- If the second flow starts at a peak or valley of a larger absolute value than the origin of the first flow, a cycle can also be counted. In this case, however, the cycle ranges between the values of the origin of the first flow and the drop-off point (Figures 4.19(d) and (f)): closed loops 2 and 5 and finally 1 and 6 are identified and saved).

Once a cycle is counted, its data points are notionally removed from the graph and the counting process continues. Figure 4.20 shows a possible result of the whole procedure corresponding with Figure 4.18. As the whole signal is subdivided into constant sections in the beginning, the counting matrix shows the amount of occurrences of closed cycles n_i from one certain level of displacement to another. Existing cycles that do not exceed the predefined section margin are ignored.

Before this two-parameter-dependent procedure is started, the structural member's response – the sensor signal of acceleration – has to be transformed into a signal of displacement. This is realized by using the FFT algorithm, which transforms the signal from a time domain function to a frequency domain one. The obtained acceleration spectrum is converted into a displacement spectrum. A major task is to prepare the correct usage of signal filters that are dependent on the identified, remarkable frequency ranges in the response history, before deriving a time-dependent progression again (inverted FFT algorithm = IFFT) to detect real dynamic displacements due to traffic loads. It is to be asserted that this approach of deriving, for example, cantilever displacements owing to traffic loads is a structure-specific approximation, which has necessarily to be optimized by additional laser-displacement measurements combined with calibration truck crossings (varying loading and velocity) and video recordings as well (sections 4.5, 5.2.7 and 5.2.8).

4.6.2 CALCULATION OF STRESSES BY FEM

As the present lifetime calculations are performed in terms of stresses (Stress-Life Approach), FE analysis is necessary for the transition of measuring data. Cracks in the compression zones tend to arrest but are typically not structurally significant. Thus members or connections for which the stress cycle is at least partially in tension are currently required to be observed, with consideration of fatigue tasks. For welded components two different methods of fatigue life prediction are compared.

The nominal stress approach deals with stresses derived from simple beam models or from coarse mesh FEM models. The full-scale test-based curves take into account residual stresses, welding profile and imperfections in the material due to manufacturing. Stress concentrations resulting from the gross shape of the structure are also included in the nominal stress [2]. Structural or geometric (hot spot) stresses include nominal stresses, stresses from structural discontinuities and presence of attachments, but they do not include stresses occurring from the presence of welds [2].

As singularities at the weld toe are difficult to represent, FE modelling cannot directly give the actual peak stress at the weld toe. Therefore various types of stress extrapolation methods have been developed to overcome this problem (Figures 4.21 and 4.22).

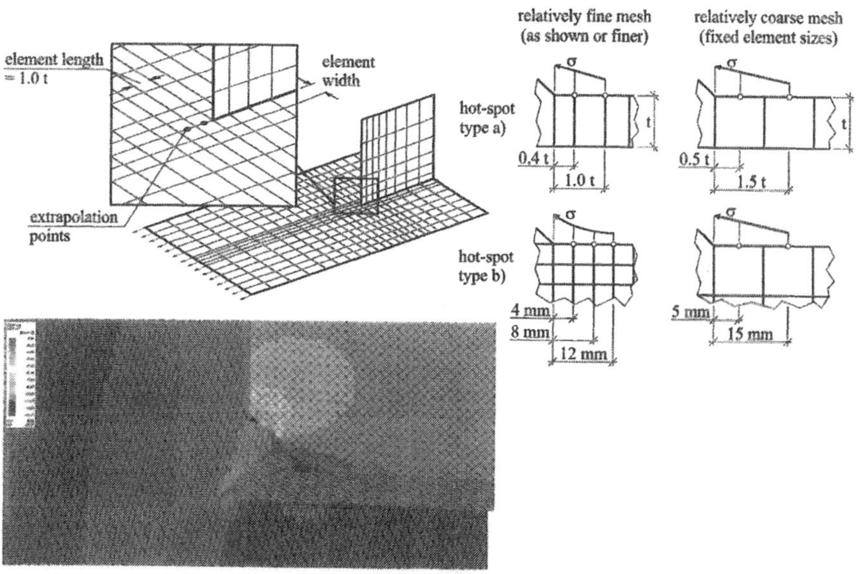

Figure 4.21 Identification of FEM-based structural (hot spot) stresses by various extrapolation procedures

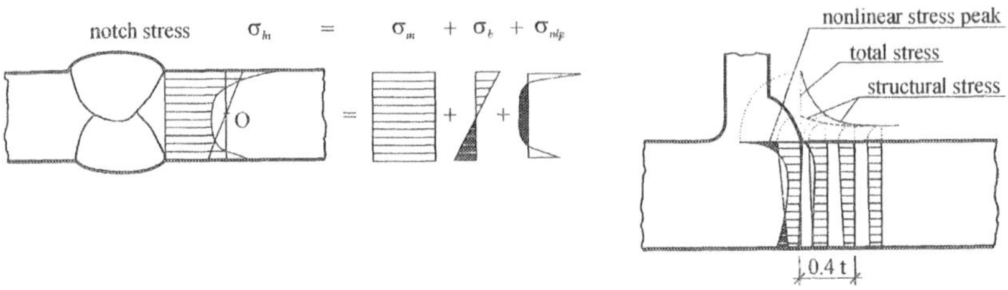

Figure 4.22 Variation in the through-thickness stress distribution approaching the weld toe [3]

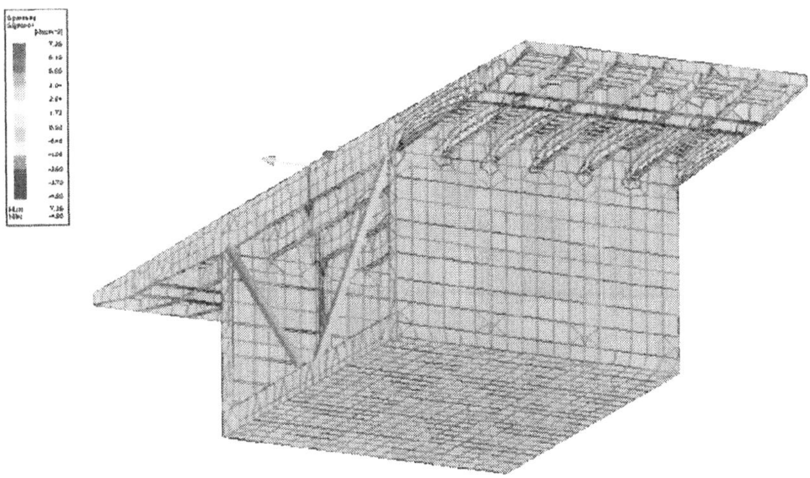

Figure 4.23 FE model of a bridge segment stressed with the unit load case along the nodes of the cantilever's outer edge (nominal stress)

Figure 4.23 shows a 9 m segment of the bridge structure representing a certain part where cantilever acceleration sensors are installed. The structure is built by shell elements modelled with a coarse mesh; therefore a refinement at the relevant hot spot areas is necessary. The hot spot *Designer's Guide* by Prof. Niemi [3] defines some directives that are of crucial importance. The maximum principle stress range within ±60° of the normal to the weld toe or the axial stress component perpendicular to the weld toe should be used for the analysis. Depending on the analysed structural detail and the used FE software, an appropriate meshing procedure is developed before a suitable formula (linear or quadratic) can be applied to derive the hot spot stresses on the plate surface by extrapolation of stresses from the calculated points in certain distances to the weld toe (e.g. Figure 4.21). Reference [4], Annex A2, includes suitable provisions of the classified S–N (stress-life) curves (Figure 4.24) that can be

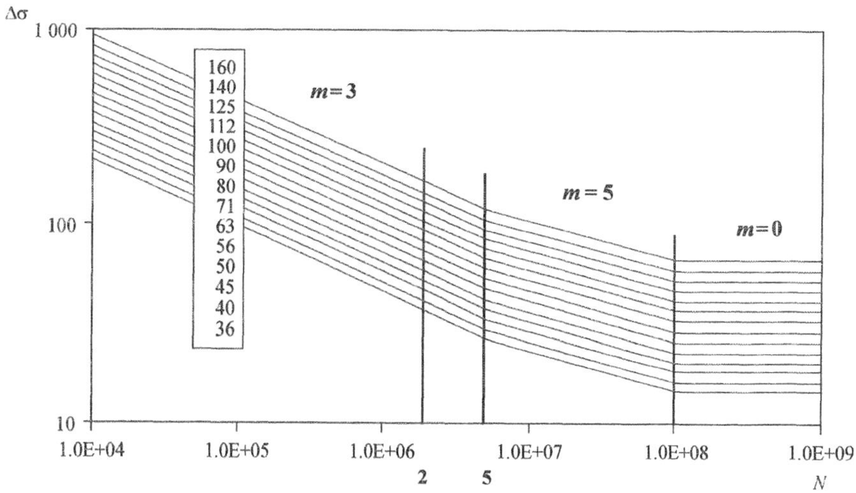

Figure 4.24 EC 3-based Wöhler curves for certain notch classes

used in connection with the geometric stress approach as an alternative to the curves published in reference [3]. A comparative lifetime calculation with nominal stresses is also to be carried out. Based on the suggestion of the authors of reference [5], these nominal stresses are identified at distance points twice the plate thickness away from the weld toe.

Figure 4.23 shows that weld geometry is neglected when using shell elements. Frequently structural intersections are chosen as a conservative solution for selecting appropriate points to represent the weld toe for stress extrapolation. It should be kept in mind that fatigue criteria were originally defined for uniaxial loading. In cases of complex loading (multi-axial, non-proportional) a fatigue criterion needs to be defined in order to apply fatigue curves obtained under uniaxial loading [6]. Even if numerous criteria have been proposed for an understanding of multi-axial fatigue, these criteria reflect only the behaviour of unwelded zones. In the current state of knowledge, the best criterion for the present work to describe the cracking of welded zones is the principal stress range, as long as the directions of the principal stress do not vary in cyclic loading cases.

Randomly influenced load–time histories lead to varying mean stresses. In cases where unwelded areas are analysed, empirical corrections (Goodman and Gerber theory [7] as an alternative) are applied. Fatigue analysis of welded assemblies has to be performed using the full stress range regardless of the mean stress during the cycle. The reason for this difference is the weld's high tensile residual stress, which reaches the yield stress of the material. The real mean stress in the material remains independent of the applied mean stress; its effect is included in the fatigue curves for welded assemblies [8]. For the same reason dead load effects are not a significant parameter in the fatigue lifetime of welded details.

4.6.3 S–N APPROACH AND DAMAGE ACCUMULATION

Fatigue analysis that compares the number of occurring loading cycles n_i on a certain stress range level $\Delta\sigma$ to an allowable number of cycles N has become a matter of course in civil engineering. Figure 4.24 shows EC 3-based S–N curves according to fatigue tests of certain constructional elements. They already include the impact of local notches due to welding, stress orientation, residual stresses, metallurgical conditions, etc. Each notch class is defined by the detail category $\Delta\sigma_c$ (belonging to two million cycles), the constant amplitude fatigue limit $\Delta\sigma_D$ (belonging to five million cycles) and the cut-off limit $\Delta\sigma_L$ (belonging to 100 million cycles).

Cases of non-periodic loading as in the present situation demand the well-known damage-accumulation concept of Palgrem-Miner, explained in reference [8]:

$$D_i = \frac{n_i}{N_i}, \qquad D_1 = \sum_{1}^{j} \frac{n_i}{N_i}, \qquad D_2 = \sum_{j+1}^{z} \frac{n_i}{N_i}, \qquad D = \sum_{1}^{z} D_i = D_1 + D_2 \tag{4.1}$$

All partial damages D_i of stress ranges corresponding to $m = 3$ are summarized in D_1, those corresponding to $m = 5$ are reflected in D_2, while stress ranges smaller than $\Delta\sigma_L$ are of no fatigue significance. Failure is predicted if $D \geq 1$ (inception of a visible crack). The non-periodic loading leads to the necessity of modelling Wöhler curves with two different slopes before reaching the cut-off limit. The decrease of slope between two and five million cycles complies with the *Modified* Miner rule [7] and represents the reduction of fatigue limit due to gradual damage in comparison to earlier stated cut-off limits. The introduced stress life approach is used for long-life applications. The fatigue threat exclusively caused by truck traffic leads to the application of high cycle fatigue theory. Stresses and strains are assumed to remain elastic. This facilitates the calculations, as a single static load case (restraint by a unit displacement) is calculated and is related to all other occurring displacements afterwards.

The following procedure describes the transition of the rainflow matrix to the damage matrix in Figure 4.25 for every analysed detail. Depending on the measured time-frame a rainflow matrix is created and extrapolated to a period of one year. After having assigned the area of interest to a certain notch class, the consequence of every element in the matrix in terms of stresses can be determined due to the unit load case. These elements cause partial damages D_i, accumulating to a total damage D per year (equation (4.1)). To obtain a better

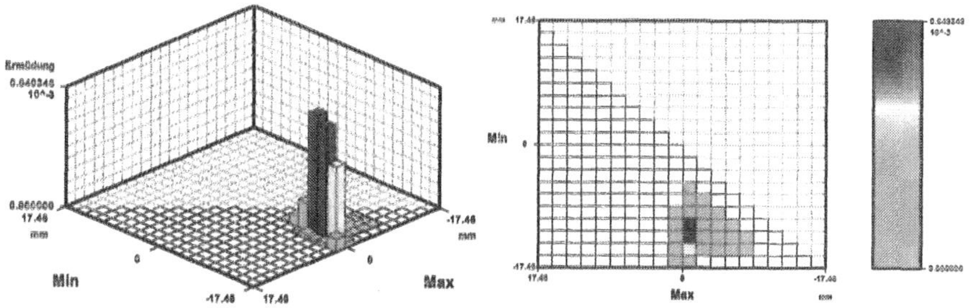

Figure 4.25 Axonometric projection of the damage matrix and its relating ground view

understanding of every element's damage relevance r, the following equation is used to transform Figure 4.20 into Figure 4.25:

$$r = \frac{D_i}{D} \times 100\ (\%) \tag{4.2}$$

4.6.4 REMAINING SERVICE LIFETIME BY MEANS OF EXISTING TRAFFIC DATA AND ADDITIONAL FORWARD AND BACKWARD EXTRAPOLATION

The assessment of an analysed detail is carried out with a damage matrix calculated for measuring time of a whole year. This matrix includes the so-called 'damage per year effect'. Detailed knowledge about the progression of the prevailing traffic [9–11] from the very beginning up to the present and the implementation of published future trend studies for the next ten years can be used for an extrapolation of the measured impact for the whole lifetime of the structure. Figure 4.26 shows the increase in freight traffic volume at the Europa Bridge. According to reference [9], traffic volume in 2003 increased to an amount of 381% in relation to 1964 and is expected to grow by 2.9% per year until 2015 [10]. To derive every considered year's damage matrix affected by the variation in traffic volume, the fatigue analysis demands an approximate uniform adaptation of the number of occurrences for all elements of the derived rainflow matrix.

Figure 4.27 shows the increase in the effective amount of transported goods compared to a notional calculated tonnage per truck. The calculations showed that this truck weight in 2003 increased to an amount of 393% in relation to 1964 and is assumed to have already reached a kind of maximum [10]. This means that a further increase in transported goods is likely to be a consequence of the still growing traffic volume. An approximate adaptation for fatigue analysis due to the variation of the notional truck weight is realized by scaling up or down those rainflow matrix parameters representing certain levels of the observed member's displacement.

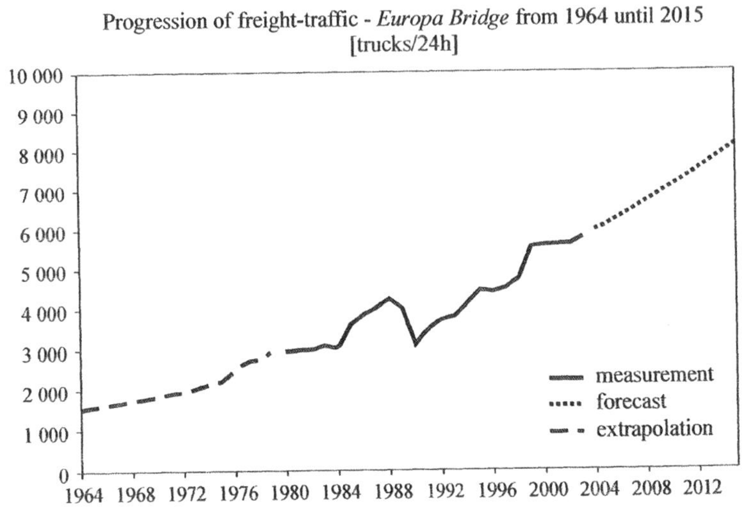

Figure 4.26 Quantitative development of traffic volume at the Europa Bridge from 1964 until 2015 (motor vehicles/day)

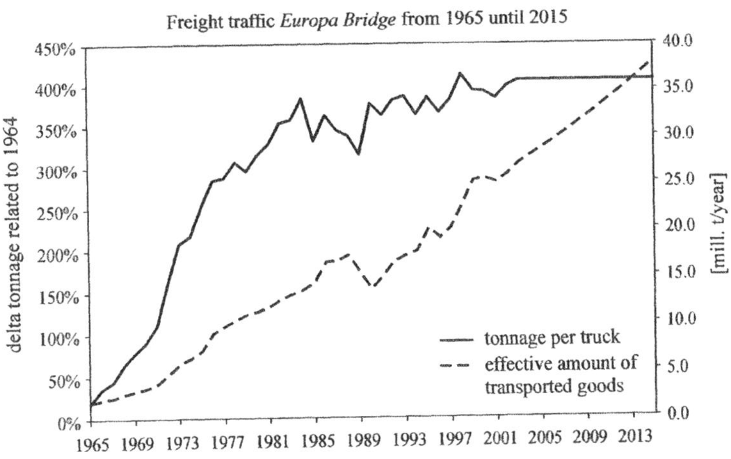

Figure 4.27 Trend of the total freight traffic on the Brenner route compared to the cargo per heavy goods vehicle

4.6.5 CONCLUSIONS AND FUTURE WORK

Up to now application of Palgrem-Miner's damage accumulation theory for research of the consequences of randomly induced traffic loads based on a permanent measuring system has been shown. The investigation's results will be improved in progressive stages, the longer the observation period lasts. This will happen using the current extrapolation procedures for the rainflow matrices presented in this section, which will be based on statistical procedures inspired by reference [7].

The accumulation of all calculated damage matrices from the very beginning of the bridge's existence up to the present leads to the remaining capacity of loading cycles for the analysed detail. As a superior conclusion in addition to a quantitative estimation of the service lifetime, another key figure, FR (fatigue relevance), is also derived. It separates the randomly occurring traffic (ADTV = average daily traffic volume) from fatigue-relevant loading cycles n_i (registered by sensors and taken from the Damage Matrix):

$$\mathrm{FR} = \frac{\sum n_i}{\mathrm{ADTV}} \tag{4.3}$$

Normally standardized stress–life curves are based on graphs derived from experimentally obtained mean values minus two standard deviations. This means that serviceability limit states are defined as exceeding 2.3% probability of failure calculated with normal distribution. As this research work tries to encourage *in situ* measurements to be taken instead of presenting 'design situations', it aspires to analyse the consequence of statistical scatter using the normal distribution, lognormal distribution (Figure 4.28) and Weibull distribution, which has already been demonstrated in diverse recently published works [12].

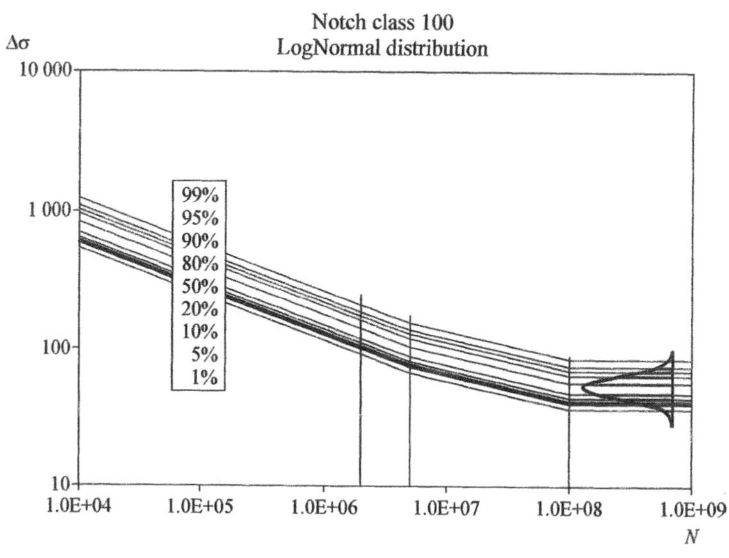

Figure 4.28 Statistical scatter of S–N curves

The fatigue process can be subdivided into two phases: initiation (= development and early growth) and propagation (= growth of a crack to failure). Up to now the stress life approach dealing with structural (hot spot) Wöhler curves given for $N = 10^4$ and higher has been discussed, as it is best suited for truck traffic, causing high cycle fatigue. It does not distinguish between initiation and propagation phases, but deals with the total lifetime. Linear elastic material behaviour can be assumed, as the structural hot spot stress range should not exceed twice the yield strength of the material [3].

Fatigue is understood as a serviceability limit state for bridges since the occurring fatigue cracks have not resulted directly in structural failure. Phenomena such as redundancy and ductility have usually prevented steel bridges from catastrophic collapse. This is the reason why preparation for sporadic but possible loading actions demanding the plastic strain life method are required, which corresponds with a number of cycles being too small to be remarkable for high cycle fatigue applications. Stress life and strain life methods are often viewed as 'crack-initiation' approaches.

The crack growth approach ignores the crack-initiation phase and is based on the fact that the analysed component is cracked before cycling begins. Further analysis observes and predicts material resistance to 'subcritical' crack growth until catastrophic fracture (Figure 4.29). Combined with the strain life

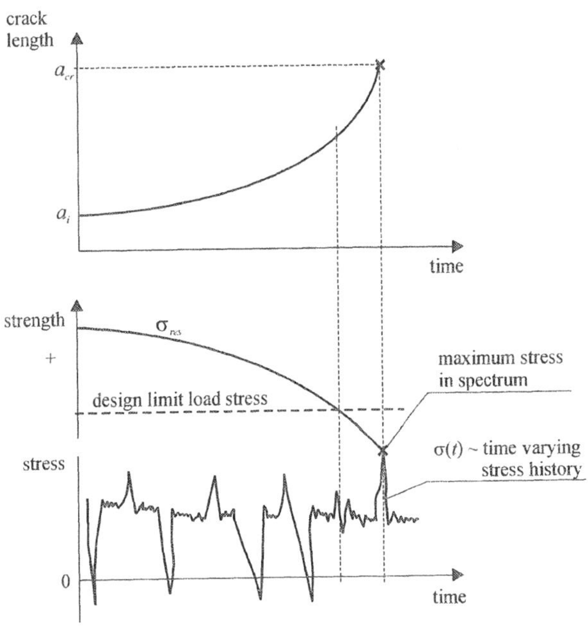

Figure 4.29 Effect of crack damage on structural integrity [13]

method this approach can also be used to predict the total fatigue life of a structural component.

REFERENCES

1. Naubereit, H. and Weihert, J. (1999) *Einführung in die Ermüdungsfestigkeit*, Carl Hanser Verlag, München-Wien.
2. Simonsen, B.C. (2001) *Procedure for Calculating Hot Spot Stresses in Aluminium Constructions*. Department of Naval Architecture and Offshore Engineering, Technical University of Denmark.
3. Niemi, E. (2001) *Structural Stress Approach to Fatigue Analysis of Welded Components – Designer's Guide*. IIW doc. XIII-1819-00/XV1090-01, Department of Mechanical Engineering, Lappeenranta University of Technology, Lappeenranta.
4. prEN 1993-1-9: 2002 *Fatigue Strength of Steel Structures*, CEN, Brussels.
5. *Ermittlung von Dauerschwingfestigkeitskennwerten für die Bemessung von geschweißten Al-Bauteilverbindungen auf der Grundlage örtlicher Strukturbeanspruchungen – Abschlussbericht* (2002) Institut für Schweißtechnik der Technischen Universität, Braunschweig.
6. *Pressure Components Fatigue Design in the Framework of Directive 97/23/EC on Pressure Equipment – Work Package 6 – Final Report* (2001) Centre Technique des Industries Mechaniques, Mulhouse, France.
7. Haibach, E. (2002) *Betriebsfestigkeit – Verfahren und Daten zur Bauteilberechnung*, VDI-Verlag, Düsseldorf.
8. Ramberger, G. (1998) *Stahlbau*, Manz-Verlag, Wien.
9. *Verkehrsentwicklung in Tirol – Bericht* (2003) Amt der Tiroler Landesregierung-Abteilung Gesamtverkehrsplanung, Innsbruck.
10. *Verkehrsprognose 2015 – vorläufige Ergebnisse hochrangiges Straßennetz Österreich* (2000) BMVIT-Abteilung II/A/1, Wien.
11. *Alpinfo 1984–2002* (2004) Federal Office for Spatial Development, Suisse, http://www.are.admin.ch/are/en/verkehr/alpinfo/.
12. IACS (International Association of Classification Societies) Recommendation 1999, No. 56, *Fatigue Assessment of Ship Structures*, July 1999.
13. ESDEP – European Steel Design Education Program. *WG12 Fatigue*, Lecture Notes, Katholieke Universiteit Leuven, Leuven, Belgium.

5

Practical Evaluation Methods

5.1 PLAUSIBILITY OF RAW DATA

Before evaluation a plausibility check of all measured data is useful. The following criteria should be considered:

1. *Quality of the signal.* Is the signal clear, is it plausible, does it show the characteristic progress, are there discernible disturbing influences by traffic, pedestrians, rotating machines with marked frequencies? Are there any other conspicuous characteristics that must be documented?
2. *Events.* Can significant events like train passages be clearly identified in the signals? Detailed aspects are as follows. Is the scanning rate as well as the level control adequate? Can the passages per span be determined in the case of continuous beams? Are the occurring maximum values plausible?
3. *Damping.* The calculation of damping in the program is done automatically for the selected eigenfrequency by the random decrement technique (RDT). If, however, a manual evaluation is carried out, this can either be done with the so-called '$\sqrt{2}$-Method' in the frequency spectrum or with the free oscillation procedure in the displacement signal.
4. *Quality of the response spectrum.* Spectra with very dominant eigenfrequencies are generally regarded as being rather unfavourable because resonance for a certain frequency is easily obtained. Broad-band spectra are more favourable with regard to energy distribution but their evaluation is considerably more difficult.
5. *Identification of the eigenfrequencies.* In the process of the evaluation it is essential to assess whether the eigenfrequencies can be clearly identified and the spectra show a progress characteristic for the structure.

Ambient Vibration Monitoring H. Wenzel and D. Pichler

6. *Residual frequencies.* If the spectrum shows frequencies that cannot be attributed to the structural response, they need to be excluded from further evaluation.
7. *Spectra in transverse and longitudinal directions.* During the assessment issues 4 to 6 are to be considered correspondingly.
8. *Position of the neutral axis.* A displacement of the neutral axis during measurement shows whether changes in the position of the sensor have occurred (defect, system displacement). The complete information content has yet to be analysed.
9. *Displacements in transverse and longitudinal directions (bearing).* With this examination the function of the bearings can be checked. The relation of the displacements in the centre of the structure to the cross-section of the bearing has to be examined.
10. *Amplitude of the eigenfrequencies.* The amplitude is a measure of the energy content of the eigenfrequencies. An assessment with regard to resonance phenomena is possible using this energy consideration.
11. *Transmission of vibrations from neighbouring structures.* The transmission is carried out via the subsoil and the foundation of the structure. This type of excitation results in a very clear response spectrum in particular for railway bridges.
12. *Deformations of the structure.* The deformations can be determined by laser measurements; the doubly integrated displacement signal can be calibrated by this. The ratio of the static to the dynamic flexure results in the dynamic factor.
13. *System identification.* The modal parameters eigenfrequency, mode shape, damping behaviour and vibration intensity are extracted from the measurements.
14. *Comparison of spectra of equal structures.* Structures with the same design should basically show the same dynamic characteristics. Eventual deviations should be analysed and the influence of different marginal conditions (e.g. superimposed load) must be considered. This examination is particularly suitable for the quality control of structures with equal design.
15. *Optical observations.* Observations or conspicuous damages are to be recorded during the measurement and should be considered for assessment.

5.2 AVM ANALYSIS

5.2.1 RECORDING

The measuring and evaluation method BRIMOS® developed by VCE is based on recording the dynamic characteristic of structures by ambient acceleration measurements by means of highly sensitive sensors. By extracting the modal parameters eigenfrequencies, mode shapes, damping as well as vibration intensities from the measuring results and comparing them with the computer-produced calculation models, statements on the actual load-bearing behaviour, the maintenance condition and forecasts for the expected future development of the structure are possible. For this purpose several steps have to be carried out:

1. Establishment of an FE Model based on the existing plans for inspections of bridge structures.
2. Establishment of the eigenfrequencies and mode shapes for the structure with an FE model by applying a dynamic analysis (natural vibrations).
3. Determination of the check points for ambient acceleration measurements based on the results of analytical tests.
4. Measurements of the vibration behaviour by means of highly sensitive acceleration sensors in established characteristic points under ambient excitation of structures and for single events (truck passage).
5. Representation of the measuring results, assessment of the quality of the individual measurements and the FFT for the determination of the response spectra for every check point.
6. Smoothing of the raw spectra, calculation of the averaged normalized power spectral density (ANPSD) and reading of the eigenfrequencies.
7. Comparison of the eigenfrequencies read from the spectra with the values calculated by the FE model, possible adaptations of the model and interpretation of deviations.
8. Determination of the mode shapes from the measurements, comparison with the calculated forms and interpretation of deviations.
9. Calculation of the material damping parameters from ambient vibration measurements by means of the RDT; assessment of the results based on tests already carried out (experience, classification according to BRIMOS®) and also on damping parameters known from literature.
10. Calculation of the system damping parameters from vibration measurements of passages and evaluation analogous to the RDT calculation.
11. Assessment of the magnitude of the acceleration acting on the structure and comparison to bridges already measured. A quantitative estimation of utilization is possible by means of a calculation with a modified FE model.
12. Representation of vibration intensity, expressed by the vibration amplitude for certain eigenfrequencies. The boundary lines of vibration intensity specify ranges for different classes of damage probability.

Items 1 to 8 can be summarized under the term 'system identification' with BRIMOS®. Items 9 to 12 represent extensions and adaptations of the dynamic measuring and assessment system for bridges for the special questions.

The dynamic measuring system BRIMOS® supplies data that can be used for the assessment of the load-bearing safety of bridge structures. The aim is to assess the safety versus the failure of the structure. The total safety of a structure is composed of several partial safety features. The latter concern both the influence and the resistance side and are summed up in five groups:

1. The load level considers the changes in standardization from the moment of planning the structure in comparison to the present as well as possible overloading of the vehicles.
2. The dynamic factor considers the specific dynamic effect of the stresses on the structure prevailing at the location. A comparison of the dynamic factor calculated according to the standard and the measured value is carried out. Here the most unfavourable measured case is always regarded as decisive.
3. In the global vibration behaviour of the structure the factor mass, E-module and stiffness are given preference. These factors essentially result from a comparison of planned (model) versus measured data.
4. The residual safety required to cover all other factors is assumed to be 1.25. This value has not yet been reflected in the literature and should therefore be discussed.
5. The interference factor describes irregularities in the behaviour of the structure. There are no fixed rules for that; the law has to be defined again for each application.

The starting point for the assessment is the planned condition. The difference between planning and measured reality is described. The result is therefore a relative value that is only valid for the specific structure.

The starting point for this evaluation method is the deterministic definition of the bridge condition under the given conditions. A probabilistic consideration is not intended because this would require several recording processes for data independent from each other (costs). This philosophy is supported by practical and budgetary considerations and could easily come into conflict with more theoretical approaches. In cases where high measuring and evaluation expenditures are justified, this method should be replaced with a probabilistic approach.

5.2.2 DATA REDUCTION

Dynamic measurements always produce high data quantities due to the high scanning rate and the many sensors. A data reduction can be reached in different ways. On the one hand, there is the possibility of securing only those data that are interesting for the evaluation of the recording already taking place. This can be carried out, for example, in the form of trigger controls where a measurement only takes place under specifically defined boundary conditions. Such a method is advisable with regard to the assessment of vibrations or the recording of stress collectives if the acquisition system records data over a longer period. On the other hand, there is the possibility of selecting data by saving the measurements as a whole and applying a reduction algorithm afterwards.

Basically data selection can also be understood as the process of cutting out certain sequences, which represent specific events, from a single measuring file. Accordingly a data record must be sufficiently long in order to be able to filter out single events from a record with as little loss in quality as possible. In this connection it must be noted that for the execution of an FFT a power of two (2^n) is used. The optimum measuring adjustment with regard to memory efficiency and quality of the results has proved to be a setting with 33 000 points per measurement with a scanning rate of 10 ms (corresponds to 100 Hz). The file is then sufficiently long to be able to cut out single events or periods without external influences on the structure.

5.2.3 *DATA SELECTION*

The raw data resulting from the measurement are processed in such a way that an optimal evaluation is enabled. The process itself is programmed and runs automatically. The individual steps carried out are:

1. Conversion of the acceleration values from mV into g, depending on the respective sensor, and elimination or suppression of non-relevant information in the files (e.g. offsets or text passages).
2. Establishment of a signal window where the x axis automatically shows the length of the files and the y axis is automatically scaled (Figure 5.1). These data are adopted in the printing formats.
3. In the case of interfering signals the data record is 'cut' in such a way that the interfering range is not considered during evaluation. The section desired can be entered as 'beginning' and 'ending', in seconds. Changes in this section immediately cause an automatic evaluation of the fixed parameters after confirmation (Figure 5.2).
4. To reach the required data length, represented by a power of two in points, the program cuts out a signal from the defined data range, which has the length of a full power of two.
5. In the case of data records between powers the program automatically uses the next smaller power of two for evaluation (further shortening of the file is possible).

5.2.4 *FREQUENCY ANALYSIS, ANPSD (AVERAGED NORMALIZED POWER SPECTRAL DENSITY)*

For the calculation of the dynamic parameters eigenfrequencies and mode shapes a three-dimensional framework programme is used. For this purpose

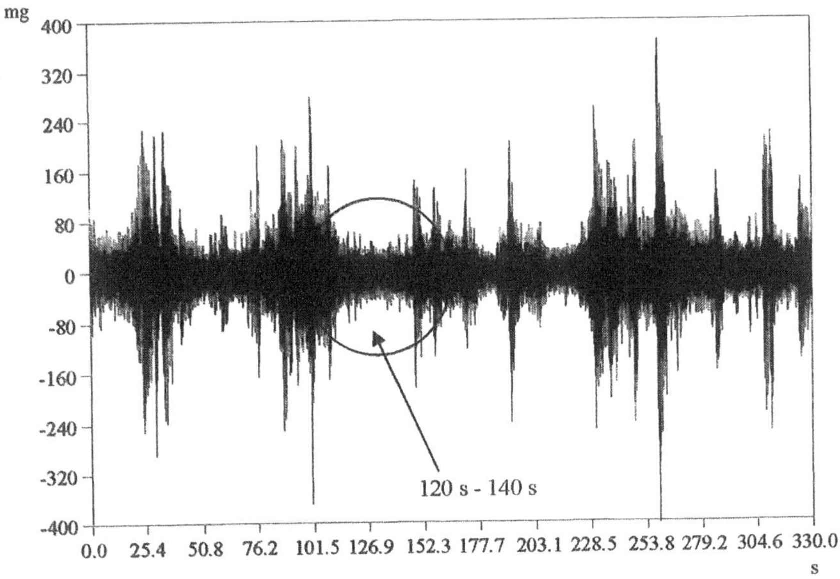

Figure 5.1 Total signal of an acceleration measurement

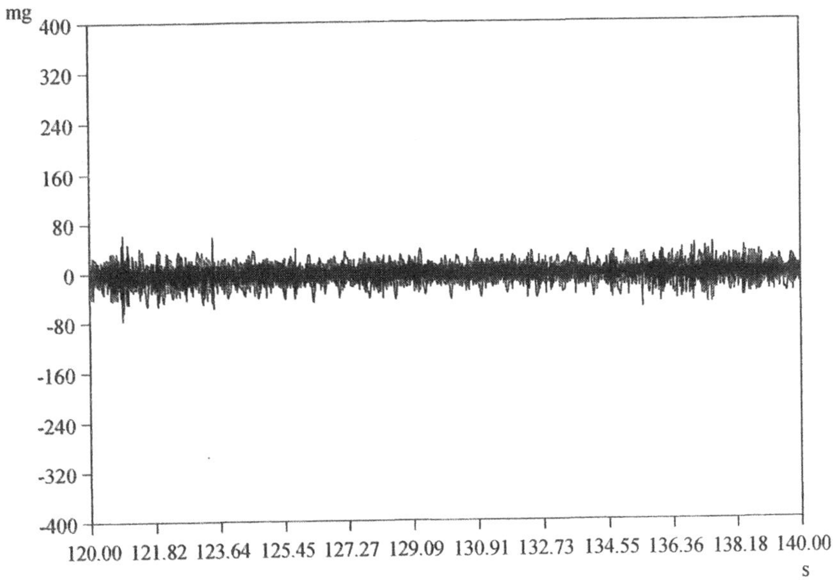

Figure 5.2 Cutting out of an ambient window

a model is established for the structure and a dynamic analysis is carried out. Figures 5.3 and 5.4 show a structural model of a bridge as defined in the RSTAB program.

Apart from a realistic recording of stiffnesses of the individual beams for the calculation model and correct consideration of the bearing conditions (degrees of freedom in the FE model), a mass distribution as faithful as possible is decisive for the quality of the results in FE examinations. Determination of the stiffnesses of individual load-bearing elements is always done on the basis of the available plan documents. The loading of the individual beams with additional masses (deck pavement, boundary beams, etc.) is also executed due to the representations in the plans by integrating the findings from inspection of the structures in the course of measuring. By comparing calculated eigenfrequencies (Figures 5.5 and 5.6) with the measuring results (given later in Figure 5.9) adaptation might be required and a second calculation might be necessary.

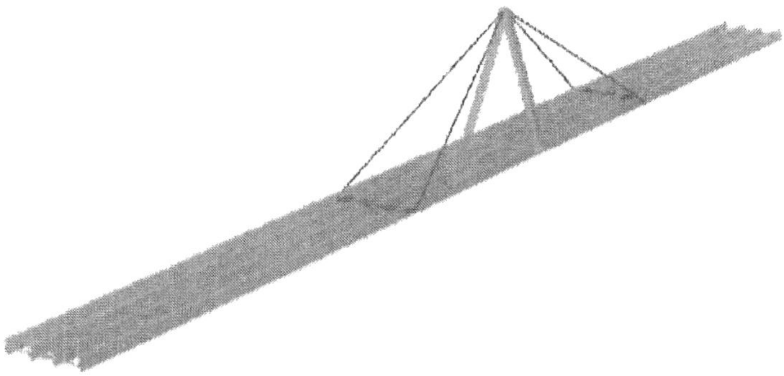

Figure 5.3 Calculation model of the Danube Bridge at Steyregg

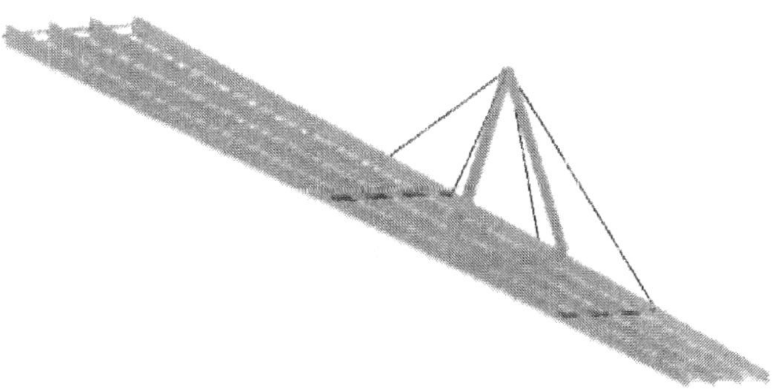

Figure 5.4 Bottom view of the FE model of the Danube Bridge at Steyregg (without composite slab)

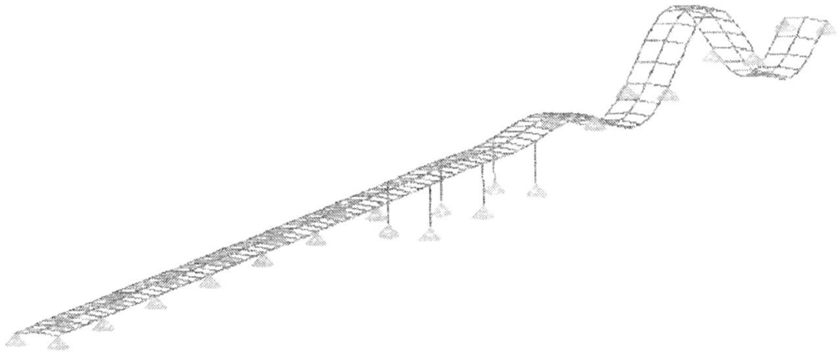

Figure 5.5 First vertical bending mode shape of the Haller Inn Bridge

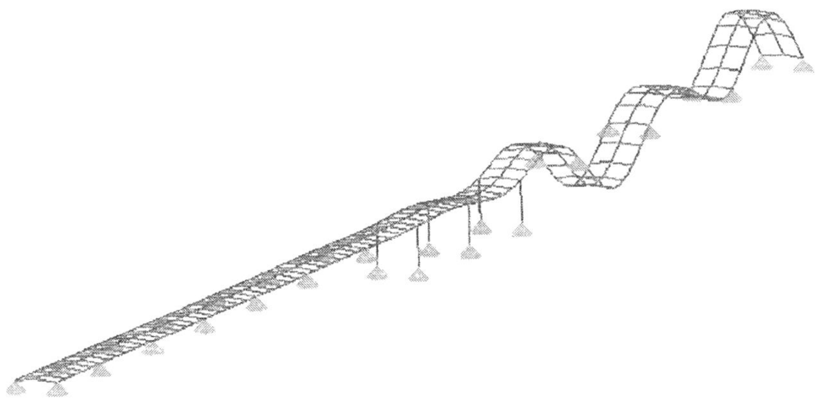

Figure 5.6 Second vertical bending mode shape of the Haller Inn Bridge

Evaluation of the data by measuring techniques is performed by means of a program. The analysis is carried out according to the following principles. The FFT calculates the spectrum of a data record according to the fast FFT algorithm. In the present implementation the classic algorithm according to Cooley and Tukey [1] in the basis 2 form is used. The FFT is a special case of the general discrete fourier transformation (DFT). In this special case, advantage is taken of the fact that the length of the data record is a power of two.

The spectral analysis is implemented separately for each channel and therefore a result comprises as many spectra as channels used for the measurement. According to the point of installation, different frequencies are dominant; e.g. the first frequency is only represented very weakly in the nodes of their respective mode shapes. Every point must therefore be individually assessed as to which spectrum to expect.

In addition to the nine vertical channels, four channels in the transverse direction and four channels in the longitudinal direction are usually measured per measurement. From this information an assessment of the load-bearing behaviour in the transverse and longitudinal directions can be gathered. In many cases the drawn-up spectra show similar characteristics to those of the vertical spectrum, which suggests the occurrence of coupled vibration forms.

The following results are described in the windows on the screen:

- vertical accelerations: four windows for the signal, raw spectra, smoothed frequencies and ANPSD (Figures 5.7 and 5.8);
- longitudinal accelerations: as before;
- transverse accelerations: as before;
- vertical displacements: four windows for the displacements, the displacement spectrum, the position of the zero line and data on the external sensor;
- longitudinal displacements: three windows for the displacements, displacement spectra and zero line;
- transverse displacements: three windows for the displacements, displacement spectra and zero line.

The smoothed and filtered representation of the spectrum expresses more than the raw spectrum in which every single frequency is represented. In the representation the values of the spectrum are transformed in such a way that the program interprets them as a curve that can be smoothed.

During the measuring period the structure is subject to static and dynamic deformations that need to be considered. For this purpose the position of the

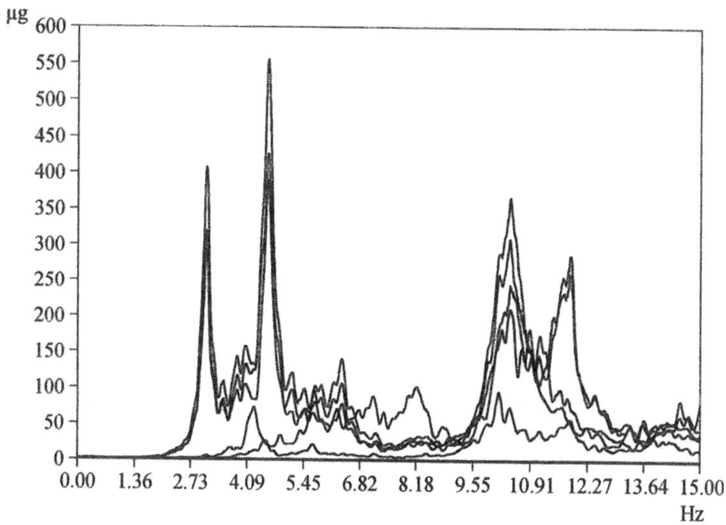

Figure 5.7 Smoothed raw spectrum of all sensors (0–15 Hz)

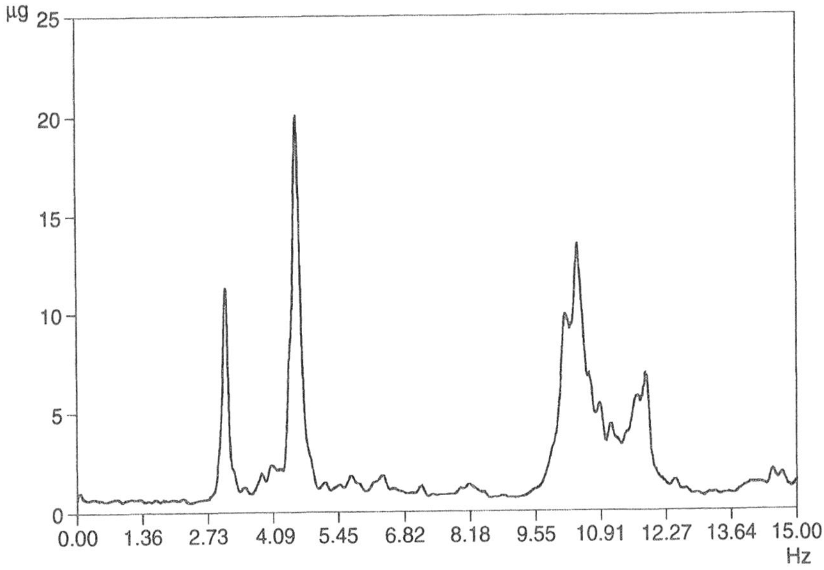

Figure 5.8 Averaged normalized power spectral density (ANPSD)

neutral axis of the sensor must be represented during the measurement. This is done in such a way that a weighted smoothing is applied to over 1000 check points. From this the chronological course of the neutral axis displacement can be recognized.

If the neutral axis displacement is very slow, it can be situated in a measuring range that is outside the sensitivity of the sensors. The procedure for determination of the neutral axis as applied here can therefore only show a trend (no actual deformations). Actual deformations have to be gauged by laser measurement equipment. As it is very difficult to adjust the neutral axis accurately because of the sensitivity of the measuring instruments, the neutral axis is situated a few millivolts above or below zero in every measurement. This is, however, considered in the evaluation by mathematical calculation.

5.2.5 *MODE SHAPES*

After conversion of the sensor signal into metres per second squared (m/s^2), calculation of the oscillation speeds can be done using a single integration. A second integration calculates the respective displacements from the speeds in the sensor positions. The integration proceeds by partial integration via four check points, which results in compensation of the variations in the individually measured values. The result of the calculation is oscillation (in millimetres) around the neutral axis; i.e. only the dynamic deformations are shown.

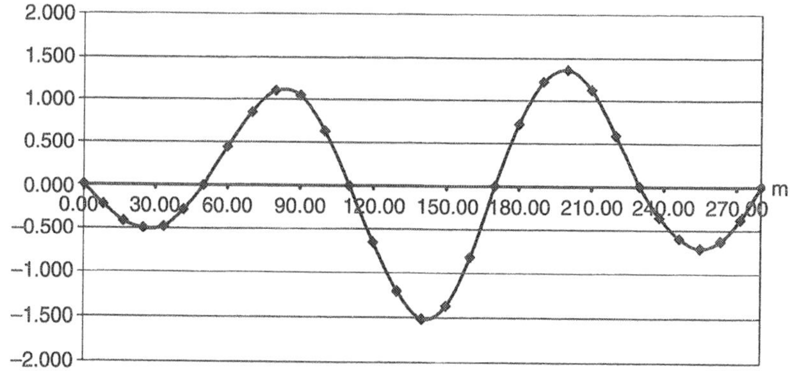

Figure 5.9 First vertical mode shape, determined from the measured data

The mode shapes are determined by comparing the respective measuring signal with the reference signal, which must also be related to the adjustment signal. In this way the relations of the deformation at the individual points can be determined. The appertaining values are determined by double integration. An examination is done by means of the optical control measurement (laser). The check points are laid out in such a way that the relevant data on the mode shapes can be derived. Therefore the mode shape characteristic must be known before the measurement in order to be able to plan a corresponding measuring layout.

For every measurement the relative value of deformations related to a reference signal of 1.00 is determined in order to represent the mode shapes. These values are taken over into an animation program where the essential geometric data of the structure are registered (e.g. the spans). By allocating these values to respective points on the structures a representation of the measured mode shape can be made (Figure 5.9). This mode shape can be compared with the mathematical mode shape. Furthermore, it is possible to import the measured mode shapes into an animation program, which enables a comparison to be made of calculated and measured mode shapes by evaluation of so-called 'MAC factors'. Here high values of the MAC factor (near 1.0) show a correlation between analytically and experimentally determined mode shapes.

5.2.6 DAMPING

5.2.6.1 General Information

One of the most important factors for the assessment of structures is damping. In the signals measured by the AVM damping is implicitly contained but is disturbed by numerous interferences from traffic and other influences. A procedure for the elimination of forced influences is required.

A similar problem arises during the determination of damping in the wind-tunnel, where the forced parts from the wind have to be eliminated so that only system parts remain. The technology used in this process is called the RDT and could be used for the AVM in an adapted form.

5.2.6.2 Principle of RDT

Under the assumption that the stimulation of the structures is white noise, the modal parameters can be derived from the measured response spectrum by correlation functions. The RDT functions according to the principle of averaging discrete time segments of a record, which follow a certain trigger function. The process is mathematically simple; the difficulty only lies in the choice of the trigger function.

5.2.6.3 Practical Implementation

As a practical procedure for the AVM program the following process is suitable:

- recording of signals with a minimum of 33 000 points;
- FFT analysis for the creation of a representative spectrum;
- selection of the relevant eigenfrequencies;
- cutting out of a window from the response spectrum by means of low-pass and high-pass filters (at least fifth order) (Figure 5.10);
- establishment of a conditioned signal by inverse FFT;
- determination of root mean square (RMS) (dynamic average);
- selection of the window length (three to five periods);
- determination of the trigger function for the start of the window;
- calculation of all windows;
- summing of all windows;
- averaging of the result;
- curve adaptation over the points obtained;
- representation of the conditioned signal;
- calculation of damping from the conditioned signal (Figure 5.10).

Installation into BRIMOS® has been possible since Version 4.0. A similar algorithm is also contained in the program MATHCAD. Studies and research works in this field were performed at the University of Aalborg (Denmark) directly based on the research of the NASA studies from the 1960s.

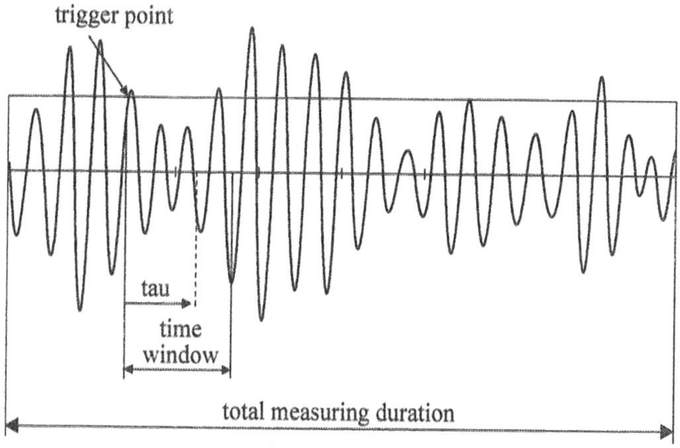

Figure 5.10 Damping window

5.2.7 *DEFORMATIONS*

Deformations cannot be directly determined by the measuring system – they implicitly exist in the data of the acceleration record. It is possible to obtain the speed signals from the acceleration signals by integration in the time range and by a further integration step, the vibration displacements.

It has to be considered, however, that the displacements determined by this procedure include the dynamic share of the structure during stress (vibration). The static deformation of the structure, which takes place, for example, during the passage of a vehicle, cannot be determined by this procedure. The static deformations can, however, be calculated from the acceleration signals by using the neutral axis around which the dynamic signals oscillate. The sensor experiences a movement reflected in the measured data during deformation of a structure. As this is a longer event, this deformation is determined during the smoothing of an acceleration signal. The neutral axis therefore represents the average value of the acceleration signal in the period observed. Direct deformations can be measured by a laser, which represents the static deformation during single event (Figure 5.11). This laser signal can be used as the reference signal for the displacement signal calculated from the acceleration.

Figures 5.12 and 5.13 show the restoring movement of a structure of Vienna's Süd-Ost-Tangente in the area of St Marx, which goes back to its original position after the release of a decisive stress. The movement was visible in the two independent systems during the deformation measurement with laser as well as during the measurement taken with the acceleration sensors.

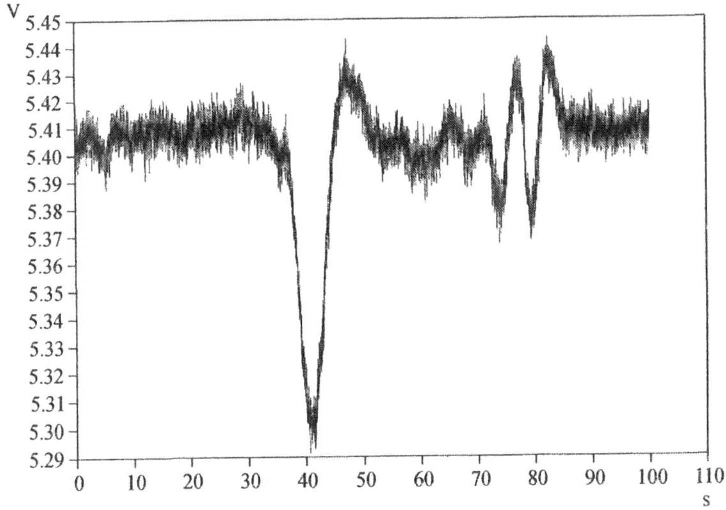

Figure 5.11 Laser signal of a truck test passage in St Marx

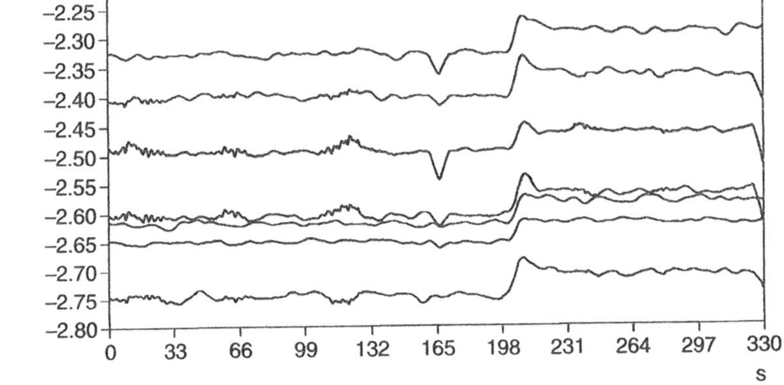

Figure 5.12 Displacement of the neutral axis due to structure displacement (basis: acceleration sensors)

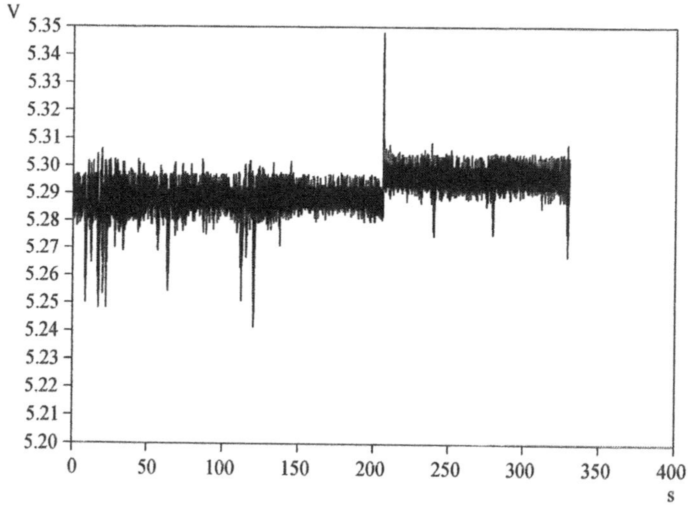

Figure 5.13 Displacement of the neutral axis (laser measurement)

5.2.8 *VIBRATION COEFFICIENTS*

The vibration coefficients for the system can be determined on the basis of the measured signals. Basically the raw deformation signals are compared to the smoothed and filtered vibrations (Figure 5.14). During the evaluation of

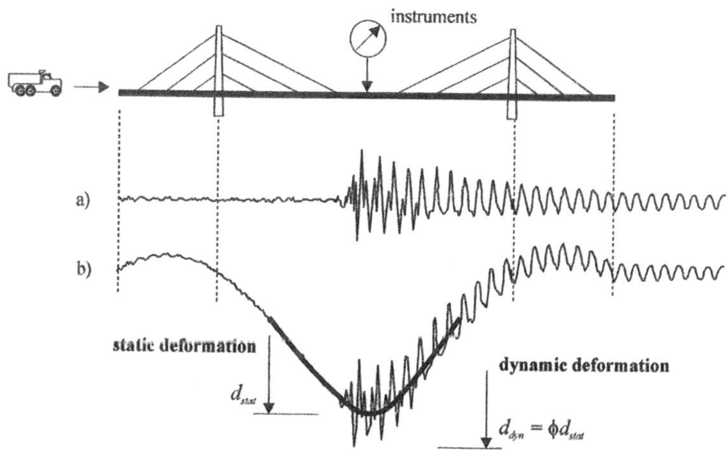

a) Relative vertical oscillation at mid-span caused by the passage of the truck.
b) Dynamic influence line of the passage of the truck. Determination of the dynamic amplification factor.

Figure 5.14 Static and dynamic deformations

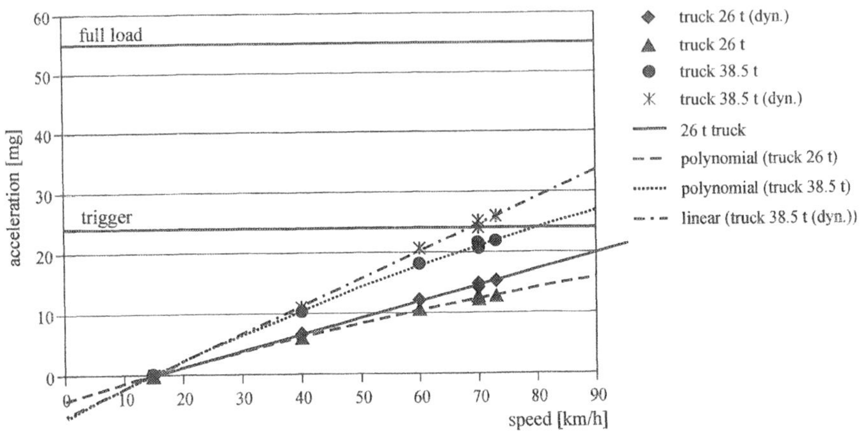

Figure 5.15 Correlation speed, signal and weight

passage events it is striking that apart from vehicle weight the passage speed plays an essential role. The evaluation algorithm can, however, not be transferred from one bridge to another. If the method is calibrated, the evaluation offers reliable information, as shown in Figure 5.15.

5.2.9 COUNTING OF EVENTS

From the dynamic characteristic all single events can be filtered out, which cause a reaction of stress in the bridge. This does not coincide with the number of truck passages as the stress can be very different. If this information is gathered over a long period of time, secured load collectives can be drawn up and the actual degree of utilization of the structure is determined (Figure 5.16).

Every stress in a structure is documented as its dynamic response. It is therefore only necessary to define the corresponding load stages and to register any exceeding defined limit values. In order to secure the results, a calibration with a known vehicle is required. By noting the passages at different speeds the influence of this parameter can be considered. Overtaking manoeuvres as well as tailgating vehicles can clearly distort the results when the weight of both vehicles is determined. For the stress, however, only the scale is interesting and not its trigger.

In order to get an impression of the triggers it has proved successful to install video monitoring where alarm pictures record when a critical value is exceeded. This makes it possible to identify overloaded vehicles (Figures 5.17 and 5.18).

Monday - Friday

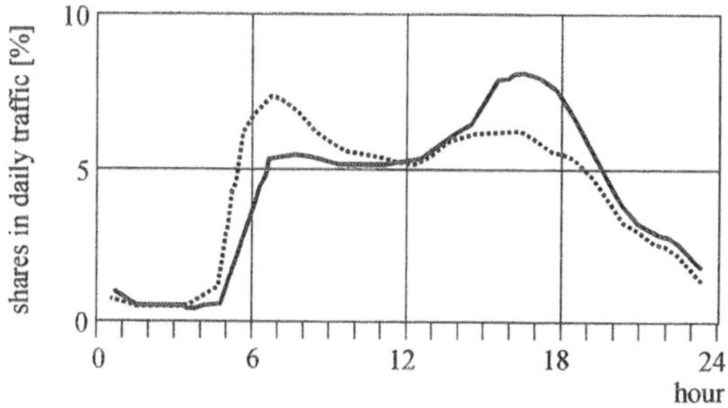

Sundays and holidays

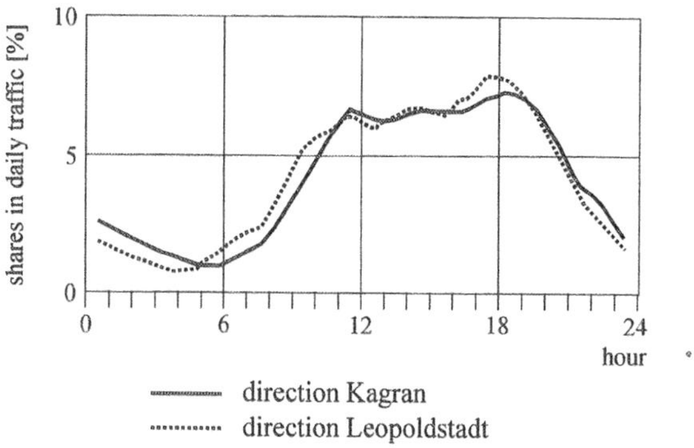

Chronological distribution on 27 January 1999

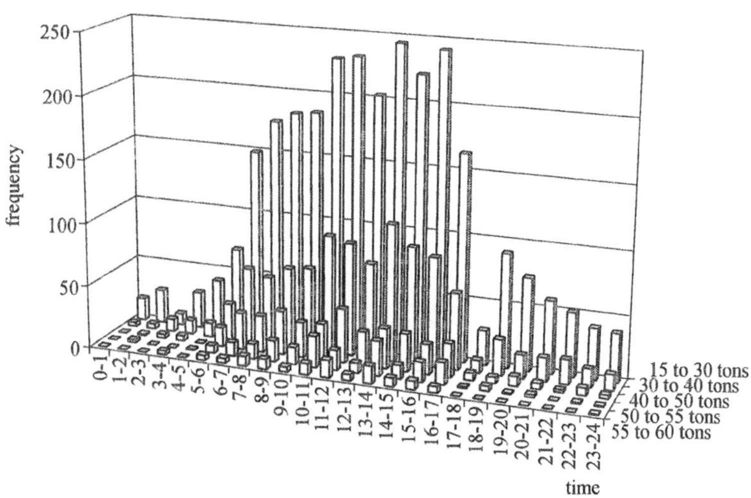

Figure 5.16 Daily traffic course in St Marx

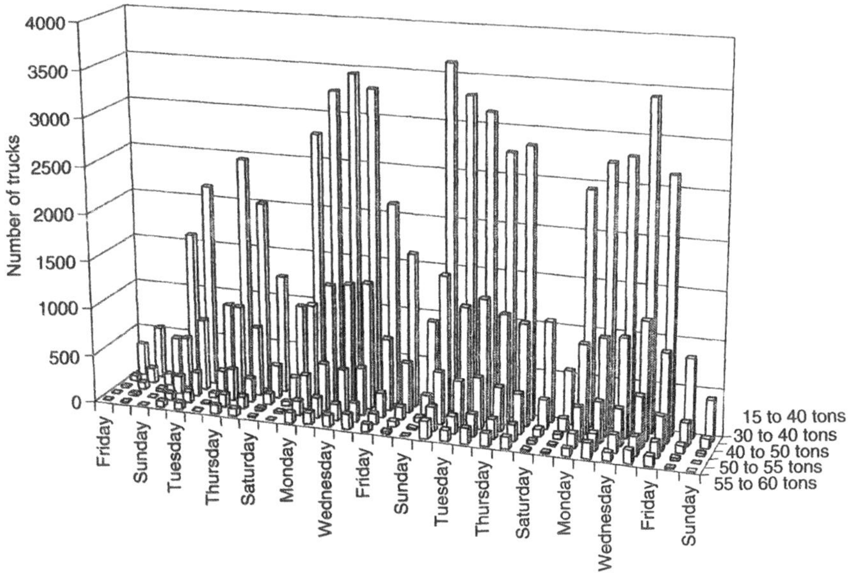

Figure 5.17 Weekly and monthly course of events

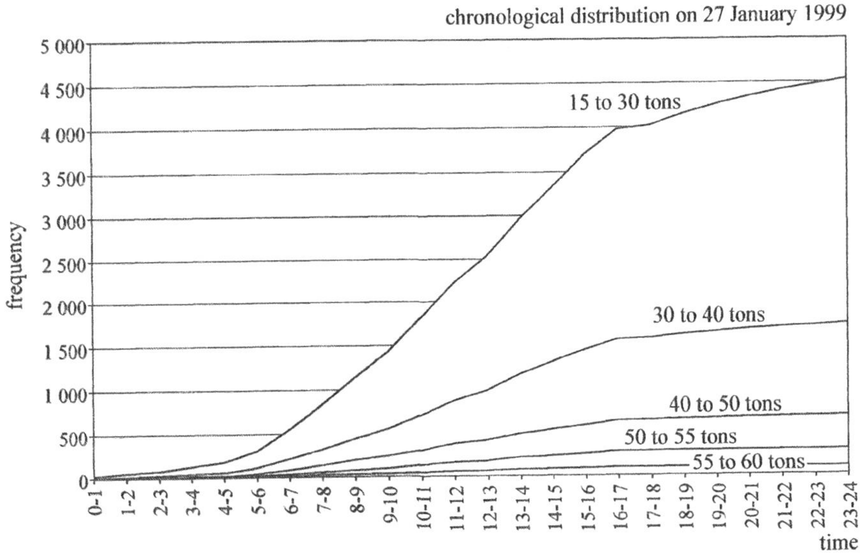

Figure 5.18 Sum lines for one day

5.3 STOCHASTIC SUBSPACE IDENTIFICATION METHOD

Contributed by Guido De Roeck, Bart Peeters and Anne Teughels, K.U. Leuven, Belgium

Structural Health Monitoring (SHM) covers the global non-destructive methods that evaluate the physical condition of structures based on vibration measurements. Vibration-based damage detection relies upon the fact that a local stiffness change affects the global dynamic characteristics of the structure.

In vibration-based health monitoring, a large amount of measurement data is generated. The amount of data is compressed by estimating an experimental modal model of the structure, consisting of eigenfrequencies, damping ratios, mode shapes and modal participation factors. The process of finding the modal model from the vibration data is called system identification [2].

In vibration-based damage detection techniques, the identification of damage is based on changes in the modal model.

5.3.1 THE STOCHASTIC SUBSPACE IDENTIFICATION (SSI) METHOD

Several models of vibrating structures exist, from models that are close to physical reality to general dynamic models that are useful in system identification. Examples of these model types are FE models of civil engineering structures, state-space models originating from electrical engineering and modal models initially developed in mechanical engineering.

System identification starts by adopting a certain model that is believed to represent the system. Next, values are assigned to the parameters of the model so as to match the measurements. Stochastic system identification methods estimate the parameters of stochastic models by using output-only data [2–4]. The methods can be divided according to the type of data that they require: frequency domain spectral data, covariances or raw time data. Accordingly, they evolve from picking the peaks of spectral densities to subspace methods that make extensive use of concepts from numerical linear algebra.

In a civil engineering context, the civil structures (e.g. bridges, towers, etc.) are the systems; the estimation of the modal parameters is the particular type of identification and stochastic means that the structure is excited by an unmeasurable input force and that only output measurements (e.g. accelerations) are available. It is assumed that the input corresponds to white noise.

The time domain data-driven stochastic methods identify models directly from the response time signals. The data-driven stochastic subspace identification (SSI) method cancels out the (uncorrelated) noise by projecting the row space of future outputs into the row space of past outputs. The idea behind this projection is that it retains all the information in the past that is useful to predict the future. Robust numerical techniques from linear algebra – such as QR factorization (an algorithm for matrix decomposition into a Q part and an R part), singular value decomposition and least squares – are used in the further processing of the data in order to solve the identification problem. The principles of a non-steady-state Kalman filter are applied for the identification of a state-space model [2–4].

Once the parametric model is identified and available, the modal parameters can then be derived easily from the model matrices.

In practice, civil structures are frequently excited by ambient forces (such as wind, traffic, etc.) or impact loads (coming from a hammer or a drop weight). The main advantage of ambient sources is the fact that the bridges can remain operational, which avoids the costs that would evolve from putting them out of use. On the contrary, artificial excitation by a shaker is not very cost-effective, since a very powerful shaker is necessary to excite the heavy structure and additional manpower is needed to install it. Furthermore, if a structure has low-frequency (below 1 Hz) modes, it may be difficult to excite it with a shaker, whereas this is generally no problem for a drop weight or ambient sources. The high-frequency modes, on the other hand, are not always well excited by ambient sources. If mass-normalized mode shapes are required, ambient excitation cannot be used. To obtain the correct scaling of the mode shapes, the applied force has to be known.

5.3.2 APPLICATION TO BRIDGE Z24

The SSI method is applied to bridge Z24 in Switzerland in order to extract the modal data of the bridge from ambient vibrations. The bridge Z24 is extensively instrumented and tested with the aim of providing a 'feasibility proof' for vibration-based health monitoring in civil engineering. The bridge is located in Canton Bern near Solothurn and connects Koppigen with Utzenstorf. It overpasses highway A1 between Bern and Zürich.

It is a classical post-tensioned concrete box girder bridge with a main span of 30 m and two side spans of 14 m (Figure 5.19). The overall length is 58 m. Both abutments consist of three concrete columns connected with hinges to the girder. Both intermediate supports are concrete piers clamped into the girder. All supports are rotated with respect to the longitudinal axis which yields a skew bridge. The bridge was demolished in 1998 because a new railway, adjacent to the highway, required a new bridge with a larger side span.

5.3.2.1 Experimental Data: Test Program

In the framework of the Brite EuRam Program CT96 0277 SIMCES, the bridge is progressively damaged in a number of damage scenarios [5,6] before

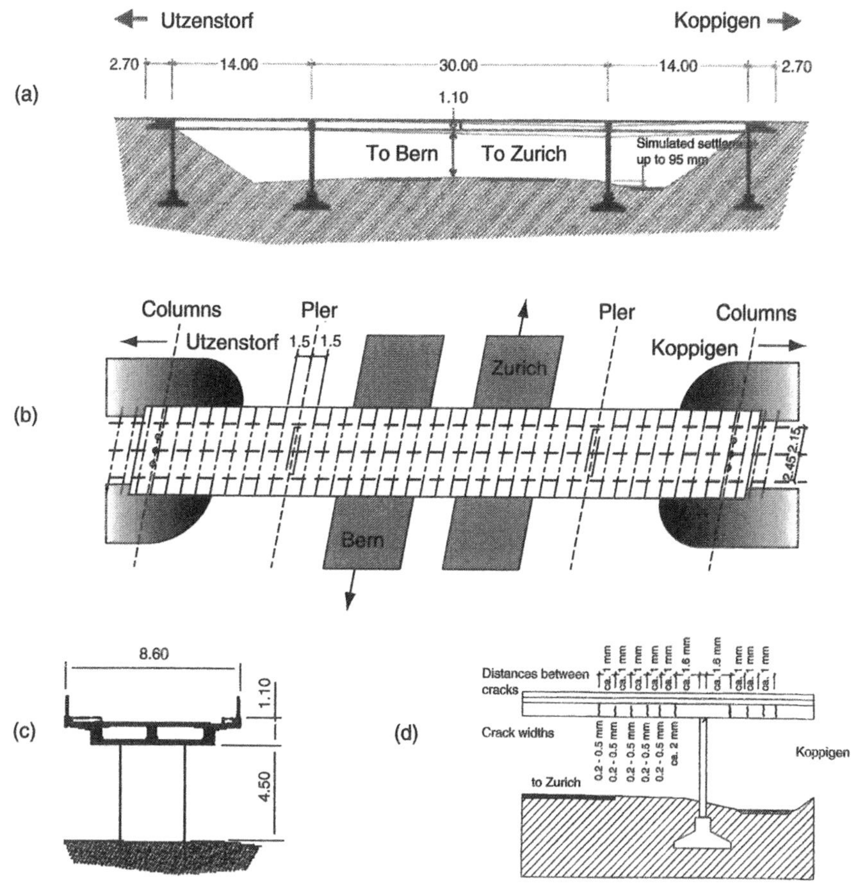

Figure 5.19 Highway bridge Z24: (a) elevation; (b) top view with measurement grid indicated; (c) cross-section; (d) crack pattern in the bridge girder, above the lowered pier

complete demolition. A full description of the damage scenarios is given by Krämer *et al.* [7]. They are listed briefly in Table 5.1.

The modal data are identified from ambient vibrations, measured at the different damage stages. The measurements are performed in operational conditions. The ambient sources acting on the bridge are highway traffic (underneath the bridge), wind and walking of the test crew in the case of low traffic density. The modal data, corresponding to different damage stages of the bridge, are identified using the SSI method.

For this purpose, accelerometers are placed on the bridge deck along three parallel measurement lines: at the centreline and along both sidelines

Table 5.1 Damage scenarios on bridge Z24 [6]

Number	Scenario	Description/simulation of real damage cause
1	First reference measurement	Initial (healthy) structure
2	Second reference measurement	After installation of the lowering system
3	Lowering of pier, 20 mm	Settlement of subsoil, erosion
4	Lowering of pier, 40 mm	
5	Lowering of pier, 80 mm	
6	Lowering of pier, 95 mm	
7	Tilt of foundation	Settlement of subsoil, erosion
8	Third reference measurement	After lifting the bridge to its initial position
9	Spalling of concrete, 12 m^2	Vehicle impact, carbonization and subsequent corrosion of the reinforcement
10	Spalling of concrete, 24 m^2	
11	Landslide at abutment	Heavy rainfall, erosion
12	Failure of concrete hinge	Chloride attack, corrosion
13	Failure of anchor heads I	Corrosion, overstress
14	Failure of anchor heads II	
15	Rupture of tendons I	Erroneous or forgotten injection of tendon tubes, chloride influence
16	Rupture of tendons II	
17	Rupture of tendons III	

(Figure 5.19(b)). Nine measurement set-ups are used to identify the mode shapes. The mode shapes are obtained by glueing the parts that are identified in each set-up using four reference channels.

The damage scenario that will be considered in the application (section 5.4.2) consists in lowering one of the supporting piers (at 44 m) by 95 mm (scenario 6 in Table 5.1), inducing cracks in the bridge girder above this pier. This simulates the settlement of the pier foundation.

The eigenfrequencies and mode shapes, identified with the SSI method, are given in Figure 5.20 for the first five eigenmodes of the bridge in its initial state and after the pier settlement. The first and the fifth are pure bending modes, the third and fourth are coupled bending and torsional modes (due to the skewness of the bridge) and the second is a transversal mode. The settlement of the pier causes a change in mode shapes 3 to 5, owing to the induced cracks in the bridge girder.

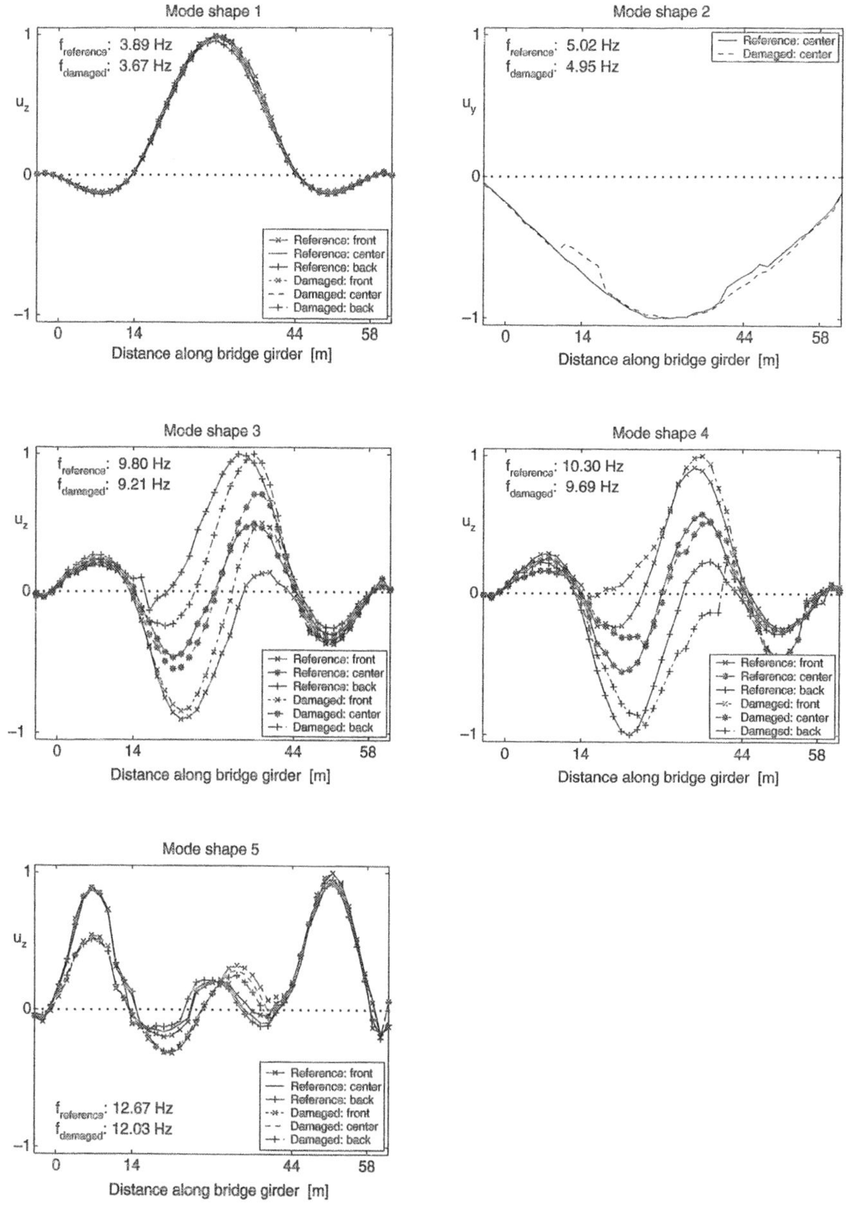

Figure 5.20 Experimental eigenfrequencies and mode shapes of bridge Z24 before and after settlement of the pier

5.4 USE OF MODAL DATA IN STRUCTURAL HEALTH MONITORING

Contributed by Guido De Roeck, Bart Peeters and Anne Teughels, K.U. Leuven, Belgium

5.4.1 *FINITE ELEMENT MODEL UPDATING METHOD*

In vibration-based damage detection techniques, the change in modal data is used as an indicator to detect and to identify the damage in the structure. An inverse problem is solved that consists in predicting the location and severity of the damage, given the structural dynamic characteristics before and after the damage.

The FE model updating method belongs to this class of damage detection techniques. The procedure consists in adapting the unknown properties of an FE model, such that differences between experimental modal data and the corresponding analytical predictions are minimized. In the case of damage identification, the structural damage is represented by a decrease in the stiffness of the individual elements; the procedure is performed in two updating processes. In the first process, the initial FE model is tuned to the undamaged structure, which is used as a reference model. In the second process, the reference FE model is updated to obtain a model that can reproduce the experimental modal data of the damaged state. The correction factors of the latter process represent the damage. In order to reduce the number of unknown parameters, damage functions are used to approximate the unknown damage pattern [8].

5.4.1.1 General FE Model Updating Procedure

The general procedure of the FE model updating method is shown in Figure 5.21. Initially, the numerical modal data are calculated using the FE model with initially estimated values for the unknown model parameters. The experimental modal data are obtained from ambient vibration tests on the structure, e.g. by using the SSI method. In an iterative process the unknown model parameters are adjusted until the discrepancies between the numerical and experimental modal data are minimized.

5.4.1.2 Objective Function

The minimization of the objective function is stated as a nonlinear least squares problem, which is defined as a sum of squared differences:

$$f(\boldsymbol{\theta}) = \frac{1}{2}\sum_{j=1}^{m}[z_j(\boldsymbol{\theta}) - \tilde{z}_j]^2 = \frac{1}{2}\sum_{j=1}^{m} r_j(\boldsymbol{\theta})^2 = \frac{1}{2}||r(\boldsymbol{\theta})||^2 \tag{5.1}$$

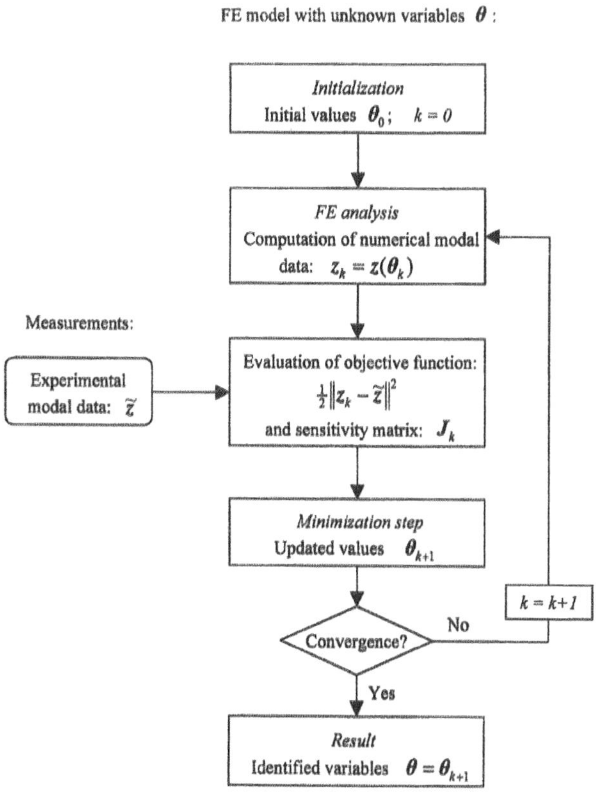

Figure 5.21 Flowchart of the FE model updating procedure

where each $z_j(\boldsymbol{\theta})$ represents an analytical modal quantity that is a nonlinear function of the optimization variables $\boldsymbol{\theta}$ ($\in \mathbb{R}^n$); $\tilde{z}$ refers to the measured value of the quantity z and $\|\cdot\|$ denotes the Euclidean norm. In order to obtain a unique solution, the number m of residuals $r_j = z_j - \tilde{z}_j$ should be greater than the number n of unknowns $\boldsymbol{\theta}$.

The residual vector $\boldsymbol{r}$:\op R^n $\rightarrow$\op R^m is split into a frequency residual vector $\boldsymbol{r}_f$ and a mode shape residual vector $\boldsymbol{r}_s$. The residuals are defined as [8]

$$\boldsymbol{r}(\boldsymbol{\theta}) = \begin{bmatrix} \boldsymbol{r}_f(\boldsymbol{\theta}) \\ \boldsymbol{r}_s(\boldsymbol{\theta}) \end{bmatrix} \quad \text{with} = \begin{cases} r_f(\boldsymbol{\theta}) & = \frac{\lambda_j(\boldsymbol{\theta}) - \tilde{\lambda}_j}{\tilde{\lambda}_j}, \quad j \in \{1, \ldots, m_f\} \\ r_s(\boldsymbol{\theta}) & = \frac{\phi_j^l(\boldsymbol{\theta})}{\phi_j^r(\boldsymbol{\theta})} - \frac{\tilde{\phi}_j^l}{\tilde{\phi}_j^r} \quad j \in \{1, \ldots, m_s\} \end{cases} \tag{5.2}$$

with eigenvalue $\lambda_j = (2\pi v_j)^2$ and eigenfrequency v_j; λ_j and ϕ_j denote the numerical eigenvalue and the mode shape vector respectively, where $\tilde{\lambda}_j$ and $\tilde{\phi}_j$ refer to the corresponding experimental values; m_f and m_s are the number of identified eigenfrequencies and mode shapes respectively, used in the updating process.

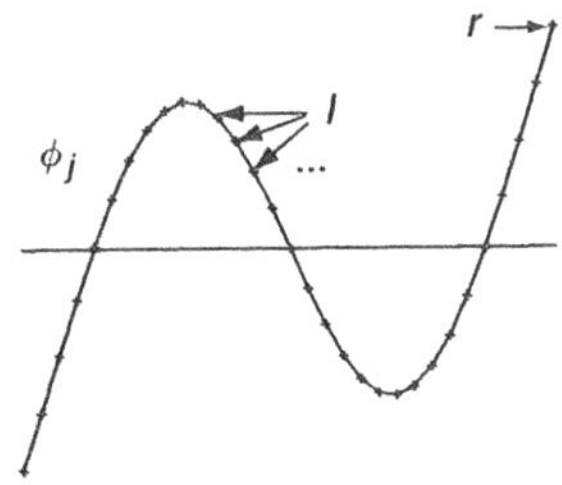

Figure 5.22 Mode shape ϕ_j with reference component r and other components l

The indices l and r of ϕ_j denote respectively an arbitrary and a reference degree of freedom of the mode shape (Figure 5.22). The total number of residuals is

$$m = m_f + \sum_{j=1}^{m_s} n\,\mathrm{DOF}_j$$

with n DOF_j representing the number of degrees of freedom (DOFs) used in mode shape ϕ_j (without counting the reference DOF).

Relative differences are taken in r_f in order to obtain a similar weight for each frequency residual. In r_s the analytical and experimental mode shapes are scaled to 1 in a reference component ϕ^r, which is a component with a large amplitude, since the experimental scaling factor is unknown if output-only data are used.

The least squares problem formulation allows the residuals to be weighted separately, corresponding to their importance and amount of noise. The weight factors influence the result only in the case of an overdetermined set of equations. Only the relative proportion of the weighting factors is important, not their absolute values. The following weighted least squares problem is solved:

$$\min \frac{1}{2} \| \boldsymbol{W}^{\frac{1}{2}} \boldsymbol{r}(\boldsymbol{\theta}) \|^2 \tag{5.3}$$

with $\boldsymbol{W}$ the weighting matrix, which is often a diagonal matrix, i.e. $\boldsymbol{W} = \mathrm{diag}(\ldots, w_j^2, \ldots)$, with w_j the weighting factor of the residual r_j.

Generally, the experimental eigenfrequencies are a good indicator of damage and can be measured quite accurately. However, it is difficult to detect zones of local damage using only eigenfrequencies. Mode shapes in their turn permit a more detailed prediction of the damage distribution, but the measurements are more noisy.

5.4.1.3 Optimization Variables $\boldsymbol{\theta}$

One or more unknown physical properties X (e.g. Young's modulus) are updated in each element e of the numerical FE model. A dimensionless correction factor a^e expresses the relative difference of the updated value of property X with respect to its initial value X_0^e in element e:

$$a^e = -\frac{X^e - X_0^e}{X_0^e} \Rightarrow X^e = X_0^e(1 - a^e) \tag{5.4}$$

The correction factors can affect one element or may be assigned to an element group. If the unknown physical property is linearly related to the stiffness matrix of the element (group), then

$$\boldsymbol{K}^e = \boldsymbol{K}_0^e(1 - a^e) \tag{5.5}$$

$$\boldsymbol{K} = \boldsymbol{K}^u + \sum_{e=1}^{n_e} \boldsymbol{K}_0^e(1 - a^e) \tag{5.6}$$

where $\boldsymbol{K}_0^e$ and $\boldsymbol{K}^e$ are the initial and updated element stiffness matrices respectively, $\boldsymbol{K}$ is the global stiffness matrix and $\boldsymbol{K}^u$ is the stiffness matrix of the element (group) whose properties remain unchanged; n_e is the number of elements (groups) that are updated.

Adjusting the model property of all the elements separately would result in a high number of updating variables $\{a^e\}$, which causes the sensitivity matrix $\boldsymbol{J}$ to become ill-conditioned for the same residual vector $\boldsymbol{r}$. Further-more, a physically meaningful optimization result is not guaranteed since neighbouring elements can be adjusted independently. Therefore, the distribution of the correction factors $\{a^e\}$, which define in their turn the distribution of the updated physical properties X over the FE model, is approximated by using global damage functions $\mathcal{N}(\boldsymbol{x},\boldsymbol{\theta})$ [8].

A damage function $\mathcal{N}(\boldsymbol{x},\boldsymbol{\theta})$ is characterized by the parameters $\boldsymbol{\theta}$ ($\in \mathbb{R}^n$) and can be generally described as a linear combination of shape functions N_i, which in their turn can be parameterized by shape parameters $\boldsymbol{t}_i$:

$$\mathcal{N}(\boldsymbol{x},\boldsymbol{\theta}) = \sum_{i=1}^{n_p} p_i N_i(\boldsymbol{x},\boldsymbol{t}_i) \quad \text{with} \begin{cases} \boldsymbol{p} = \{p_1,\ldots,p_{n_p}\}, & \boldsymbol{p} \in \mathbb{R}^{n_p} \\ \boldsymbol{t} = \{t_1,\ldots,t_{n_p}\}, & \boldsymbol{t} \in \mathbb{R}^{n_t} \end{cases} \tag{5.7}$$

Note that $\boldsymbol{\theta} = \{\boldsymbol{p},\boldsymbol{t}\}$ contains both the set of multiplication factors $\boldsymbol{p}$ for the linear combination as well as the sets of shape parameters $\boldsymbol{t}_i$ of each shape function N_i ($\boldsymbol{\theta} \in \mathbb{R}^n$, $n = n_p + n_t$); $\boldsymbol{x}$ defines the position on the FE model ($\boldsymbol{x}$: geometrical coordinates).

The actual optimization variables are the set of parameters $\boldsymbol{\theta} = \{\boldsymbol{p},\boldsymbol{t}\}$ that determine the damage function $\mathcal{N}$ $(\boldsymbol{x},\boldsymbol{\theta})$ uniquely. In order to obtain the values for the individual elements of the FE model, the damage function is discretized in the element centre, which corresponds to taking a constant correction value a^e for each updated element:

$$a^e = a(\boldsymbol{x}^e)\sum_{i=1}^{n_p} p_i N_i(\boldsymbol{x}^e, \boldsymbol{t}_i) \tag{5.8}$$

with $\boldsymbol{x}^e$ the coordinates of the centre of element e. The equivalent vector notation is

$$\boldsymbol{a}_{n_e\times 1} = [\boldsymbol{N}(\boldsymbol{t})]_{n_e\times n_p}\,\boldsymbol{p}_{n_p\times 1} \tag{5.9}$$

with $[\boldsymbol{N}(\boldsymbol{t})]$ the matrix that contains each shape function N_i, evaluated in the element centres, as a column. In this way, the number of optimization variables can be reduced considerably (if $n \ll n_e$), which is favourable for stability of the optimization process.

Furthermore, a smooth result is always obtained, which is determined by the damage function. The latter should therefore be selected appropriately such that a realistic and physically meaningful result is obtained.

An efficient damage function is a piecewise linear function (shown in Figure 5.23, left), which is obtained by combining triangular shape functions N_i that differ from zero only over a limited area of the FE model (Figure 5.23, right). The shape functions N_i themselves are not parameterized in this case (i.e. no shape parameters $\boldsymbol{t}$ have to be defined) and the global damage function is simplified to

$$\mathcal{N}(\boldsymbol{x},\boldsymbol{\theta}) = \mathcal{N}(\boldsymbol{x},\boldsymbol{p}) = \sum_{i=1}^{n} p_i N_i(\boldsymbol{x}) \tag{5.10}$$

The actual optimization variables $\boldsymbol{\theta}$ are the multiplication factors $\boldsymbol{p} \in \mathbb{R}^{n_p}$; thus $n = n_p$. A mesh of damage elements is defined on top of the mesh of finite elements, with each damage element simply consisting of a set of neighbouring finite elements (Figure 5.23, bottom right). The functions N_i are defined with respect to a node of this mesh and differ from zero only in the adjacent elements and equal zero in the other elements.

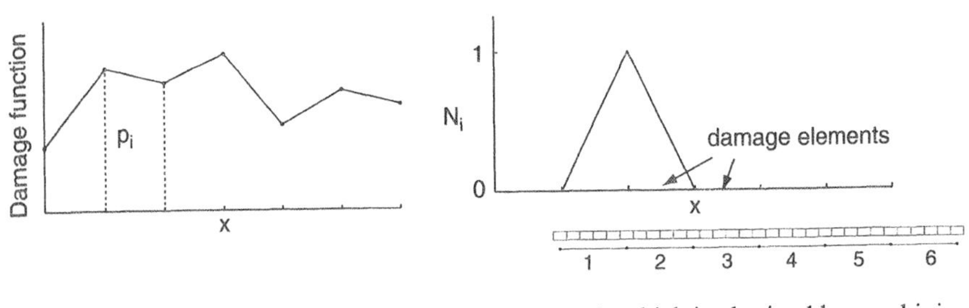

Figure 5.23 A piecewise linear damage function (x, **p**), which is obtained by combining triangular shape functions defined on a damage element mesh. It is illustrated here on a beam model

The accuracy of the updating result is determined by the coarseness of the damage element mesh and can be improved by refining the mesh, resulting in more linear pieces (damage elements) used to approximate the continuous distribution. Alternatively, higher order functions can also be added to improve the accuracy. Both methods result in more unknown parameters $\boldsymbol{\theta}$ to be identified and a good balance between both the required accuracy and the condition of the optimization problem should be maintained. The better the quality and quantity of the measurement information, the finer the mesh can be.

5.4.1.4 Optimization Algorithm: Trust Region Gauss–Newton Method

The nonlinear least squares function (equation (5.1)) is generally solved using the Gauss–Newton method, which is an iterative sensitivity-based optimization method that exploits the special structure of the least squares problem. In fact, the gradient and the Hessian of the objective function (equation (5.1)) have the following special structures:

$$\nabla f(\boldsymbol{\theta}) = \sum_{j=1}^{m} r_j(\boldsymbol{\theta})\nabla r_j(\boldsymbol{\theta}) = \boldsymbol{J}_{\boldsymbol{\theta}}(\boldsymbol{\theta})^{\mathrm{T}}\boldsymbol{r}(\boldsymbol{\theta}) \tag{5.11}$$

$$\nabla^2 f(\boldsymbol{\theta}) = \boldsymbol{J}_{\boldsymbol{\theta}}(\boldsymbol{\theta})^{\mathrm{T}}\boldsymbol{J}_{\boldsymbol{\theta}}(\boldsymbol{\theta}) + \sum_{j=1}^{m} r_j(\boldsymbol{\theta})\nabla^2 r_j(\boldsymbol{\theta}) \approx \boldsymbol{J}_{\boldsymbol{\theta}}(\boldsymbol{\theta})^{\mathrm{T}}\boldsymbol{J}_{\boldsymbol{\theta}}(\boldsymbol{\theta}) \tag{5.12}$$

with $\boldsymbol{J}_{\boldsymbol{\theta}}$ the Jacobian matrix (see subsection 5.4.1.5), containing the first partial derivatives of the residuals r_j (r_f and r_s) with respect to $\boldsymbol{\theta}$. In the Gauss–Newton method [9], the Hessian is approximated with the first-order term in equation (5.12), which is equivalent to solving the following linear least squares problem in each iteration k:

$$\min_{z} qk(z) = \frac{1}{2}||\boldsymbol{r}(\boldsymbol{\theta}_k) + \boldsymbol{J}(\boldsymbol{\theta}_k)z||^2 \qquad \text{with } \boldsymbol{\theta}_{k+1} = \boldsymbol{\theta}_k + z_k \tag{5.13}$$

where $q_k(z)$ is the quadratic model function that approximates $f(\boldsymbol{\theta})$ at the current vector $\boldsymbol{\theta}_k$; z denotes the step vector from $\boldsymbol{\theta}_k$.

The standard Gauss–Newton method is further stabilized by implementing it with the trust region strategy, which enhances the optimization process, causing it to converge. In order to prevent the iterates from taking extremely large steps in the case of an ill-conditioned sensitivity matrix, the algorithm determines in each iteration k a 'trust region' surrounding $\boldsymbol{\theta}_k$ where the model function q_k (in equation (5.13)) can be trusted. Typically, the trust region is a sphere defined by $||z|| \leq \Delta_k$, where $\Delta_k > 0$ is called the trust region radius.

A candidate for the new iterate, $\boldsymbol{\theta}_{k+1} = \boldsymbol{\theta}_k + z_k$, is then computed by approximately minimizing q_k inside the trust region. Thus equation (5.13) becomes

$$\min_z qk(z) = \frac{1}{2}\|\boldsymbol{r}(\boldsymbol{\theta}_k) + \boldsymbol{J}(\boldsymbol{\theta}_k)z\|^2 \qquad \text{such that } \|z\| \leq \Delta_k \tag{5.14}$$

If the candidate does not produce a sufficient decrease in f (i.e. the original objective function in equation (5.1)), which indicates that the model function q is an inadequate representation of f, the subproblem equation (5.14) is resolved with a smaller trust region Δ_k. Otherwise the candidate is accepted as a new iterate from which the process reiterates. Since in this case the model function is generally reliable, the trust region might be increased.

More information regarding the trust region approach can be found in references [9–11]. In MATLAB software [12] the trust region implementation is provided so that the algorithm can be performed automatically.

In FE model updating, the trust region strategy is an additional measure to improve the robustness of the updating procedure. The most effective measure to treat the ill-posedness of the inverse problem, however, is provided by the damage functions.

5.4.1.5 Sensitivity Matrix

The modal sensitivities with respect to the correction factors a^e can be calculated using the formulas of Fox and Kapoor [13]. If only stiffness parameters have to be corrected, these formulas are simplified to

$$\begin{aligned}\frac{\partial \lambda_j}{\partial a^e} &= \phi_j^{\mathrm{T}} \frac{\partial \boldsymbol{K}}{\partial a^e} \phi_j \overset{\text{equation (5.6)}}{=} -\phi_j^{\mathrm{T}} \boldsymbol{K}_{\mathrm{ref}}^e \phi_j \overset{\text{equation (5.5)}}{=} \phi_j^{\mathrm{T}} \frac{-\boldsymbol{K}^e}{1-a^e} \phi_j \\ &= -\frac{\phi_j^{\mathrm{T}} \boldsymbol{F}_j^e}{1-a^e}\end{aligned} \tag{5.15}$$

$$\begin{aligned}\frac{\partial \phi_j}{\partial a^e} &= \sum_{q=1;q\neq j}^{d} \frac{\phi_q}{\lambda_j - \lambda_q}\left(\phi_q^{\mathrm{T}} \frac{\partial \boldsymbol{K}}{\partial a^e} \phi_j\right) \overset{\text{equation (5.6)}}{=} \sum_{q=1;q\neq j}^{d} \frac{\phi_q}{\lambda_j - \lambda_q}\left(-\phi_q^{\mathrm{T}} \boldsymbol{K}_0^e \phi_j\right) \\ &= \sum_{q=1;q\neq j}^{d} \frac{\phi_q}{\lambda_j - \lambda_q}\left(\frac{-\phi_q^{\mathrm{T}} \boldsymbol{F}_j^e}{1-a^e}\right)\end{aligned} \tag{5.16}$$

where $\boldsymbol{F}_j^e$ represents the forces at the nodes of element e corresponding to mode shape ϕ_j. Instead of the complete base (in equation 5.16) d denotes the analytical model order) a truncated base is used, which should be high enough in view of the condition of the sensitivity matrix. In the sensitivity expressions above the (analytical) mode shapes ϕ are mass-normalized. In the residual vector (equation (5.2)), however, both analytical and experimental mode shapes are scaled to one in the reference node.

The modal sensitivities (equations (5.15) and (5.16)) are substituted in the sensitivities of the actual residuals r_j:

$$\frac{\partial r_f}{\partial a^e} = \frac{1}{\tilde{\lambda}_j} \frac{\partial \lambda_j}{\partial a^e} \tag{5.17}$$

$$\frac{\partial r_s}{\partial a^e} = \frac{1}{\phi_j^r} \frac{\partial \phi_j^l}{\partial a^e} - \frac{\phi_j^l}{(\phi_j^r)^2} \frac{\partial \phi_j^r}{\partial a^e} \tag{5.18}$$

which are used to calculate the sensitivity matrix $\boldsymbol{J}_a$ of the residual vector $\boldsymbol{r}$ with respect to the correction factors $\boldsymbol{a}$. Based on the mutual dependency between $\boldsymbol{a}$ and $\boldsymbol{\theta}$, expressed by the global damage function $\mathcal{N}(\boldsymbol{x}, \boldsymbol{\theta})$, each component of the sensitivity matrix $\boldsymbol{J}_\theta$ is calculated as

$$\frac{\partial r_j}{\partial \theta_i} = \sum_{e=1}^{n_e} \frac{\partial r_j}{\partial a^e} \frac{\partial a^e}{\partial \theta_i} = \sum_{e=1}^{n_e} \frac{\partial r_j}{\partial a^e} \frac{\partial \mathcal{N}(\boldsymbol{x}^e)}{\partial \theta_i} \overset{\text{equation (5.10)}}{=} \sum_{e=1}^{n_e} \frac{\partial r_j}{\partial a^e} N_i(\boldsymbol{x}^e) \tag{5.19}$$

in which equations (5.17) and (5.18) have to be filled in. The right-hand side of equation (19) is valid for the piecewise linear damage function. Equivalently, in matrix notation,

$$[J_\theta]_{m \times n} = [J_a]_{m \times n_e} \left[\frac{\partial \mathcal{N}}{\partial \theta}\right]_{n_e \times n} \overset{\text{equation (5.10)}}{=} [J_a]_{m \times n_e} [N]_{n_e \times n} \tag{5.20}$$

where $\boldsymbol{J}_\theta$ and $\boldsymbol{J}_a$ are the sensitivity matrices with respect to the optimization variables $\boldsymbol{\theta}$ and to the element correction factors $\boldsymbol{a}$ respectively. $\boldsymbol{N}$ is the matrix containing the shape functions as columns.

5.4.2 *APPLICATION TO BRIDGE Z24*

The FE model updating technique is applied to identify the damage of the highway bridge Z24 in Switzerland, caused by lowering one of the intermediate piers by over 95 mm. The first five identified eigenmodes are used for the updating. The aim is to detect, localize and quantify the damage pattern by adjusting the stiffness of the bridge girder.

5.4.2.1 FE Model

The bridge is modelled with a beam model (6 DOFs in each node) in ANSYS [14] (Figure 5.24). Equivalent values for the cross-sectional area, the bending and torsional moment of inertia of the box section of the main girder (Figure 5.19(c)) are calculated. The girder has higher stiffness values above the supporting piers (see Figures 5.26 (a) and (b) later) because of an increased thickness of the bottom and top slabs.

To model the girder 82 beam elements are used. The principal axes of the piers are rotated to model the skewness of the bridge. The width of the piers is taken into account by means of specific constraint equations. Mass elements are used for the cross girders and foundations. The concrete is considered to be homogeneous, with an initial value for Young's modulus of $E_0 = 37.5$ GPa and $G_0 = 16$ GPa for the shear modulus.

In order to account for the influence of the soil, springs are included at the pier and column foundations, at the end abutments and around the columns (Figure 5.24). The initial values of the soil stiffness are taken as: $K_{v,p} = 180 \times 10^6$ N/m^3, $K_{h,p} = 210 \times 10^6$ N/m^3 (under piers, at $x = 14$ and 44 m); $K_{v,c} = K_{h,c} = 100 \times 10^6$ N/m^3 (under columns, at $x = 0$ and 58 m); $K_{v,a} = 180 \times 10^6$ N/m^3, $K_{h,a} = 200 \times 10^6$ N/m^3 (at abutments); and $K_{v,ac} = K_{h,ac} = 100 \times 10^6$ N/m^3 (around columns).

The eigenfrequencies and MAC values calculated with the initial FE model are listed in Table 5.2.

5.4.2.2 Correction Factors

Two updating processes are performed in order to model the reference and the damaged state of the bridge respectively. The bending as well as the torsional stiffness of the beam elements of the girder are updated since the identified modes contain coupled bending-torsion modes as well as pure bending. They

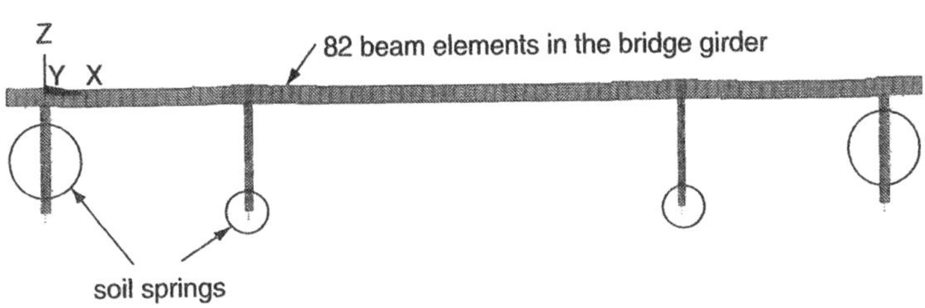

Figure 5.24 FE model of bridge Z24, where 82 beam elements are used to model the girder. The soil springs at the supports are indicated

Table 5.2 Experimental, initial and updated eigenfrequencies and MAC values for the undamaged and damaged bridge Z24. [MAC: $|\phi^T\tilde{\phi}|^2/(\phi^T\phi)(\tilde{\phi}^T\tilde{\phi})$]

Mode	Undamaged			Damaged		
	Experiment	FE model		Experiment	FE model	
		Initial	Updated		Reference	Updated
	Eigenfrequencies (Hz)			Eigenfrequencies (Hz)		
1	3.89	3.73	3.87	3.67	3.87	3.65
2	5.02	5.14	5.03	4.95	5.03	4.86
3	9.80	9.64	9.72	9.21	9.72	9.12
4	10.30	10.25	10.31	9.69	10.31	9.73
5	12.67	12.52	12.81	12.03	12.81	12.16
		MAC values (%)			MAC values (%)	
1		99.95	99.95		99.85	99.89
2		99.80	99.82		97.16	97.39
3		94.42	98.99		89.02	98.17
4		96.85	99.44		84.66	93.30
5		96.18	96.61		86.61	97.56

are adjusted by correcting Young's modulus and the shear modulus, E and G respectively:

$$a_E^e = -\frac{E^e - E_{\text{ref}}^e}{E_{\text{ref}}^e} \Rightarrow E^e = E_{\text{ref}}^e(1 - a_E^e) \tag{5.21}$$

$$a_G^e = -\frac{G^e - G_{\text{ref}}^e}{G_{\text{ref}}^e} \Rightarrow G^e = G_{\text{ref}}^e(1 - a_G^e) \tag{5.22}$$

Both properties can be updated separately by using the appropriate DOFs in equations (5.15) and (5.16), namely $\{u_x, u_y, u_z, \text{rot}_y, \text{rot}_z\}$ for the bending stiffness and $\{\text{rot}_{xx}\}$ for the torsional stiffness. The reference values, E_{ref}^e and G_{ref}^e, are the initial FE values in the first updating process, while in the second updating process they are the identified values from the first updating process.

In the first updating process, in addition the vertical soil stiffness under the supporting piers, $K_{v,p}$, and the horizontal soil stiffness under the end abutments, $K_{h,a}$, are updated. The former mainly influences the second and the fifth modes (transversal and bending), the latter only the second mode. The other soil stiffness values do not influence the considered modal data. Since the soil springs are not altered by the damage application, they are not updated in the second updating process.

5.4.2.3 Damage Function

The bridge girder is subdivided into eight damage elements: four damage elements in the mid span and two damage elements in each side span (Figure 5.25). Two (identical) piecewise linear damage functions are used for identifying the bending and the torsional stiffness distribution respectively.

In the first updating process the optimization problem contains 16 ($=2\times 7+2$) optimization variables, corresponding to the multiplication factors of both damage functions, $\boldsymbol{p}_{E,i}$ and $\boldsymbol{p}_{G,i}$ (2×7), and the two correction factors for the soil springs. In the second process only 14 ($=2\times 7$) variables need to be identified.

5.4.2.4 Objective Function

The four vertical modes (bending and bending–torsion) and the transversal mode of the undamaged bridge are used to update the initial FE model to the reference undamaged state of the bridge. The latter mode is included in the process in order to identify the stiffness of the soil springs.

The residual vector (equation (5.2)) in the reference updating process contains five frequency residuals r_f and 492 mode shape residuals r_s. The vertical displacements u_z along the three measurement lines ($\pm 3\times 39$ points) and the horizontal displacements u_y along the centreline (31 points) are used for the vertical and transversal modes respectively. Only the well-measured displacements are selected. The total residual vector $\boldsymbol{r}$ contains $m=497$ residuals.

For identification of the damaged zone only the four bending modes are used, measured on the bridge after the pier settlement. The transversal mode is

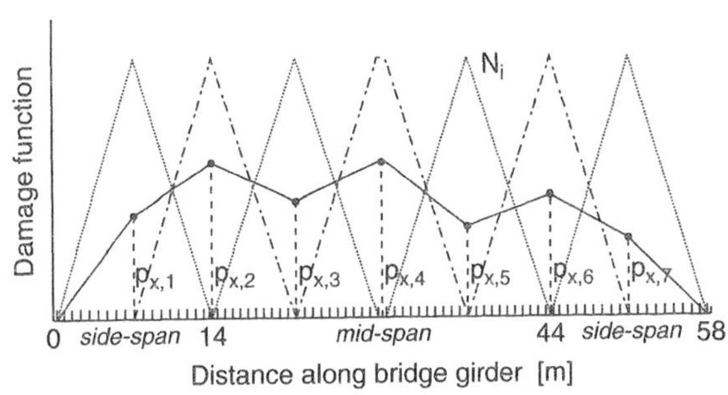

Figure 5.25 Piecewise linear damage function $\mathcal{N}(x, \boldsymbol{p})$ used to identify the distribution of both sets of correction factors, $\boldsymbol{a}_E$ and $\boldsymbol{a}_G$, for the reference and damaged states of bridge Z24 (X denotes either E or G). The bridge girder is subdivided into eight damage elements. The mesh of finite elements is also plotted on the horizontal axis

not used since the soil springs are not updated in this process. The residual vector in the second updating process contains four frequency residuals r_f and 451 mode shape residuals[1] r_s, which results in $m = 455$ residuals.

In both processes a weighting factor $w_s = \frac{1}{10}$ is applied (equation (5.3)) to the mode shape residuals.

5.4.2.5 Updating Results

The updated values of the vertical soil stiffness under the piers and the horizontal stiffness under abutments are: $K_{v,p} = 157 \times 10^6\ \text{N/m}^3$ and $K_{h,a} = 145 \times 10^6\ \text{N/m}^3$. These values are used in the FE model when identifying the damage.

The stiffness distributions of the bridge girder – for bending as well as for torsion – are plotted in Figures 5.26(a) and (b). The initial and the updated values for the reference and damaged states are shown.

The reference state is characterized by a symmetrical stiffness pattern. The initial bending stiffness is increased above both piers, at the side spans and slightly in the middle of the bridge.

In the damaged state a decrease in the girder stiffness above the pier at 44 m is clearly visible. This decrease is due to the lowering of the pier, which induced cracks in the beam girder at that location (Figure 5.1(d)). The corresponding identified damage patterns, defined by the reduction factors $\boldsymbol{a}_E$ and $\boldsymbol{a}_G$, are plotted in Figures 5.26(c) and (d). The bending and torsional stiffnesses are reduced to a maximum of 35 and 24% respectively, located in the expected cracked zone.

Some inaccuracies occur at the left side of the bridge girder (e.g. a non-physical increase in torsional stiffness) and are due to the coarseness of the damage functions, the measurement errors and the modelling assumptions. In fact a beam model is used that is not able to model the structural behaviour of the box girder exactly (no modelling of restrained warping, shear lag effects, etc.).

In a computationally more expensive calculation, a more detailed damage pattern could be obtained using an FE model with three-dimensional brick elements and a finer mesh of (three-dimensional) damage functions. The higher number of unknowns in its turn requires a larger set of noise-free experimental modal data.

5.4.2.6 Modal Data

Table 2 lists the initial and updated modal data for the undamaged as well as for the damaged bridge. In the former all five modes are used in the updating

[1] The selected displacements are plotted later in Figure 5.11 for the damaged bridge.

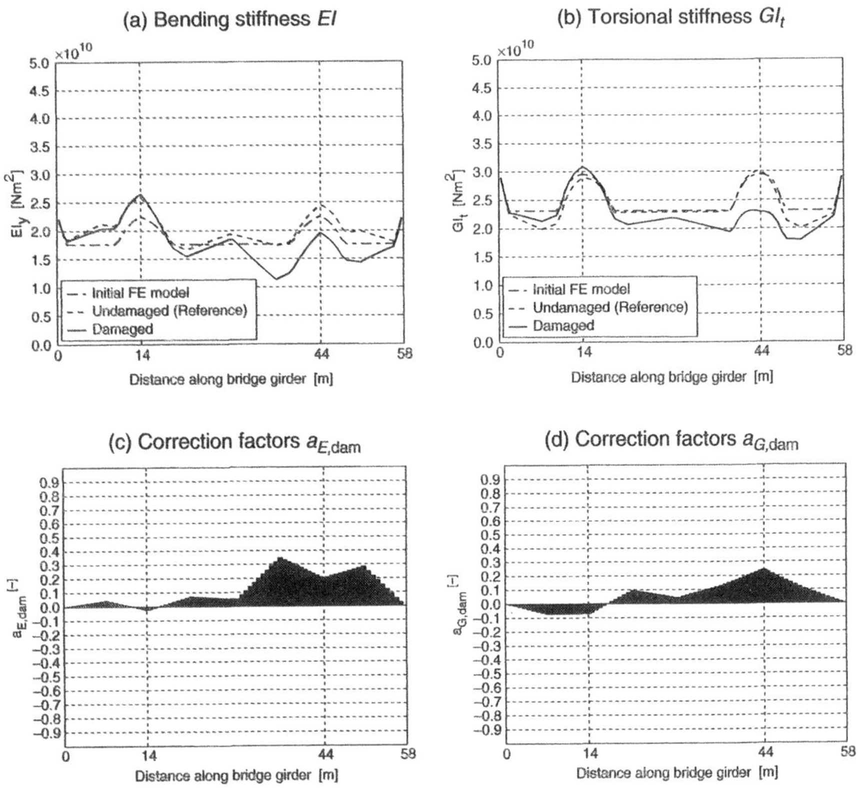

Figure 5.26 Identified parameters: (a) bending stiffness distribution $EI_{\text{dam}} = EI_{\text{ref}} (1 - a_{E,\text{dam}})$; (b) torsional stiffness distribution $GI_{t,\text{dam}} = GI_{t,\text{ref}} (1 - a_{G,\text{dam}})$, and their correction factors, (c) $\boldsymbol{a}_E$ and (d) $\boldsymbol{a}_G$, for the damaged bridge

process, whereas in the latter only the bending modes (modes 1, 3, 4 and 5) are used.

By updating the initial FE model to the reference state, the numerical and experimental eigenfrequencies correspond much better and a clear improvement for the MAC values can be observed (Figure 5.27). In particular, the correction of the soil spring stiffness reduces the discrepancy in eigenfrequency for the transversal mode.

For the damaged bridge, the correlation between the numerical and experimental eigenfrequencies is also greatly improved for the updated FE model (Figure 5.28). The updated numerical mode shapes correspond more clearly with the experimental mode shapes (Figure 5.29). The MAC value for the fourth mode shape, however, remains under 95%, which is partially due to the bad quality of the experimental data of this mode shape.

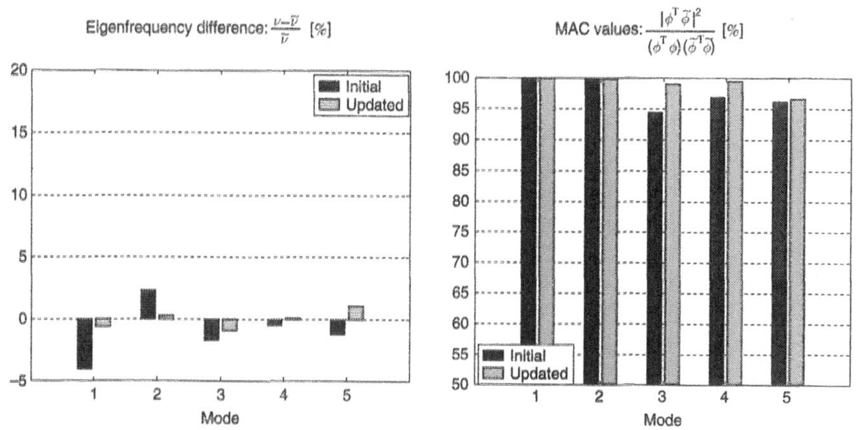

Figure 5.27 Relative eigenfrequency differences and MAC values between numerical and experimental modes for the undamaged bridge

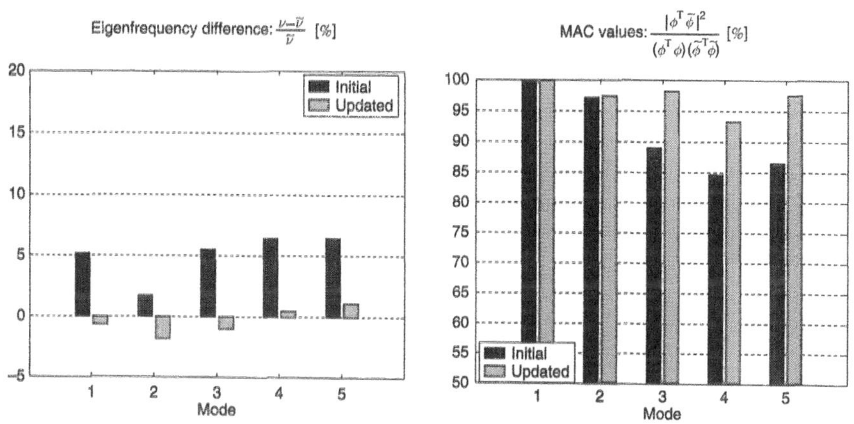

Figure 5.28 Relative eigenfrequency differences and MAC values between numerical and experimental modes for the damaged bridge. The transversal mode (mode 2) is not used in this updating process

5.4.3 CONCLUSIONS

In structural health monitoring (SHM), the aim is to assess the condition of civil structures by monitoring the changes in modal parameters. The experimental modal data are obtained from vibration measurements on the structure, which is excited dynamically. The procedure to identify the modal parameters from the vibration data is known as system identification.

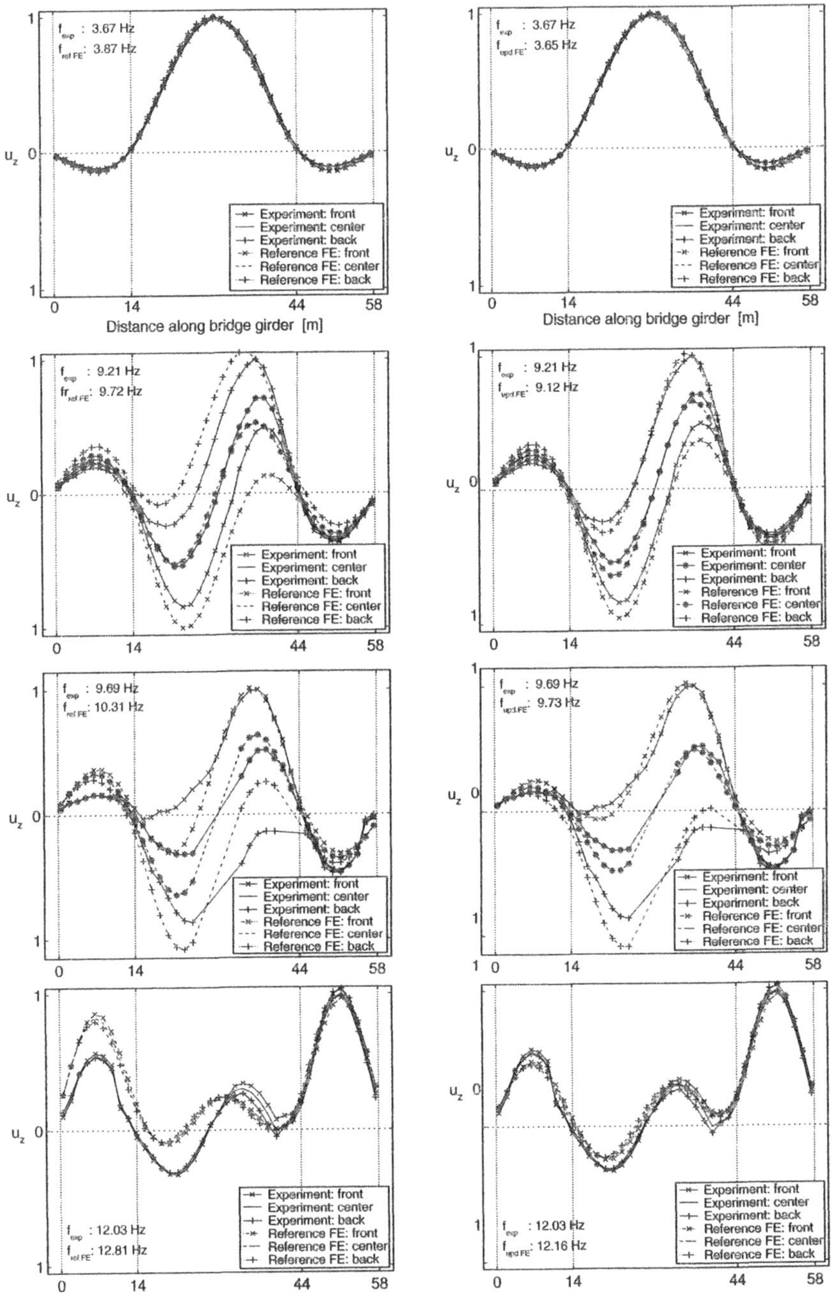

Figure 5.29 Initial (left) and updated (right) numerical modes (modes 1, 3, 4 and 5) of the damaged bridge, to be compared with the experimental values

First the section presents briefly the stochastic subspace identification (SSI) technique and illustrates it with an application to the highway bridge Z24, which has been damaged in several scenarios. The SSI method is a robust identification method that uses only time domain output data. It is assumed that the input corresponds to white noise. Frequently, ambient excitation sources are used, which makes the method very useful for the identification of (heavy) civil structures. The identified modal data can be used for a condition assessment of the structures.

The FE model updating method is an efficient vibration-based damage detection technique and is presented extensively in the section. This approach requires an analytical model of the structure. The parameters of the model that are related to damage are updated so that the dynamic characteristics of the model correspond to the measurements.

A minimization problem is solved in which the differences between the numerical and experimental modal data are minimized. Generally the eigenfrequencies and mode shapes are used for tuning the FE model. The general sensitivity-based updating method is ameliorated by the use of damage functions, in order to improve the problem condition and to ensure a physically meaningful solution. Furthermore, the optimization algorithm is stabilized by implementing it with the trust region approach.

The method is applied to the highway bridge Z24 in Switzerland. Its damage pattern, corresponding to a pier settlement, is identified using the eigenfrequencies and unscaled mode shape data, obtained from ambient vibrations by using the SSI method. The damage is represented by a reduction in bending and torsional stiffnesses of the bridge girder. For both properties a realistic damage pattern is identified.

5.5 EXTERNAL TENDONS AND STAY CABLES

5.5.1 GENERAL INFORMATION

The increasing number of external pre-stressed elements in modern structures as well as the rehabilitation or reinforcement of existing bridges with this technology raises the question of the need for a technically and economically useful check of effective cable forces or their chronological development. Principally, control of the force with a new application of the pre-stressing jack is possible, but this check is connected with a great logistic, time, and therefore also financial expenditure. Furthermore, there is the danger that unintended damage at the cables may be caused. For this reason modern, non-destructive procedures are required. Therefore methods such as the AVM are predestined since they enable a quick, flexible and safe determination of the cable force. A usual measuring interval lasts for about five and a half minutes, but in addition the time required for mounting the acceleration sensor and initializing the system has to be considered. Periods of about 15 to 20 minutes (including an evaluation of the cable force) have to be estimated for checking the cable, which is a further advantage of this system.

Up to now numerous projects have been completed using the AVM where special questions concerning cables had to be answered. Outstanding examples of this are the bridge over the Danube at Tulln, where all cables were checked for the effective cable force and the susceptibility to vibrations, as well as the Donaustadt Bridge in Vienna, where inspections due to vibration susceptibility of a cable were ordered. A check on the cable forces of external pre-stressing elements was carried out at the Mur Bridge in St Michael on the Pyhrn Motorway, at the Donnergraben Bridge during rehabilitation and at two new bridge structures in Germany, for quality control reasons.

5.5.2 *THEORETICAL BASES*

A useful, very accurate and at the same time very economical determination of the cable force can be carried out by means of the measurement of the eigenfrequencies of cable oscillations. The cables are stimulated to oscillate by traffic or other environmental (ambient) reasons. By recording the effective acceleration a subsequent conversion of the signals into frequency spectra is possible when a simple FFT is applied.

The spectrum, which represents the reaction (structural response) of the cable, very clearly shows the individual frequencies of the harmonic oscillations. As the eigenfrequencies of higher order are a multiple of the first frequency for cables without bending stiffness, it is possible to determine the cable's stiffness by analysing the deviation of the frequency course of the linear relation. The identified eigenfrequency (f) is a function of the effective cable force (N), the cable length (l), the cable mass per unit (m) and the related bending stiffness (ξ)

$$f_k = \frac{k}{2l}\left(\frac{N}{m}\right)^{\frac{1}{2}}\left[1 + \frac{2}{\xi} + \left(4 + \frac{k^2\pi^2}{2}\right) \times \frac{1}{\xi^2}\right] \qquad (5.23)$$

$$\xi = l\left(\frac{N}{EI}\right)^{\frac{1}{2}} \qquad (5.24)$$

5.5.3 *PRACTICAL IMPLEMENTATION*

For practical implementation on site, a notebook, which has to be equipped with specific software, an acquisition system (data-logger) and acceleration

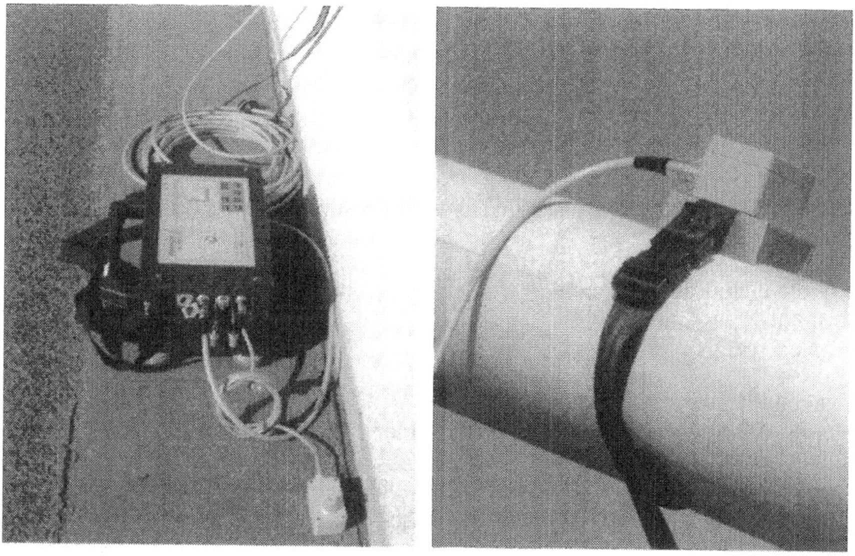

Figure 5.30 Acquisition system on the bridge deck and accelerator on the stay cable

sensors with data cables are required. The complete equipment (Figure 5.30) can be stowed in the boot of a normal motor car, which proves the economic efficiency of the method. The power supply is provided by the power supply system (230 V) or by batteries (12 V).

For precise identification of the cable force the sensor has to be placed on the cable or tendon by mounting straps and connected with the acquisition system via the data cable. After recording the measured data they are read in a special program that automatically carries out an evaluation of the eigenfrequencies. By applying the above-mentioned equation, a determination of the effective cable force can be executed very quickly. To make data independent of the environmental influence temperature a temperature sensor is connected to the acquisition system. The measuring interval usually amounts to approximately $5\frac{1}{2}$ min.

5.5.4 STATE OF THE ART

The system works without any problems when used by experts since appropriate application requires some experience. The data evaluation of a cable or tendon itself is undergoing an automation process at the moment. During the European project IMAC methods for accurate cable force determination and therefore corresponding software were developed for fast and easy use. The aim of current projects is to integrate the cable force determination in a knowledge-based system.

5.5.5 *RAIN–WIND INDUCED VIBRATIONS OF STAY CABLES*

Stay cables tend towards oscillations with big amplitudes if certain conditions are fulfilled:

- cable inclination between 20 and 30° to the horizontal;
- slight to medium rainfall;
- wind speeds of 9–12 m/s.

The amplitudes occurring in these cases are in the range of 5–6D (D = diameter of the cable). The amplitudes are dependent on the length of the cable. Frequencies of a higher order are often observed, with vibrations frequently occurring in the second and third eigenfrequencies.

At present it is only possible to predict such vibration susceptibilities of individual cables to a limited degree; therefore an observation of the cables is decisive. The following procedure has proved to be favourable:

- The cables are to be observed and possible vibrations documented.
- Susceptible cables are assessed using the AVM and their damping at the respective frequencies are determined.
- If problems are perceptible, dampers are installed.

The following measures can be taken as damping mechanisms:

- the application of an elliptical groove at the sheath for draining off the water;
- the installation of an oil damper in the form of a clip on the cable;
- the use of shock absorbers at both cable ends;
- interference cables (thin connection cables between the individual cables);
- active or semi-active damping elements.

Experience shows that only a few from a great number of cables oscillate. It is therefore not economic to take preventive measures in the form of dampers as this is very expensive compared to a subsequent rehabilitation.

5.5.6 *ASSESSMENT*

Cable vibrations caused by wind or rain were observed at many cable-stayed bridges. Examples to be mentioned are the Erasmus Bridge in Rotterdam (closure for five days) and the Maiko Nishi Bridge in Japan. On that bridge the vibration amplitudes reached several metres, which resulted in damage to the cables and anchorages. In many cases the simultaneous occurrence of wind and rain was observed. These vibrations already begin at low wind speeds, involving in most cases several mode shapes.

The AVM enables the assessment of the susceptibility of cables with regard to the two most frequent cases of cable vibrations:

- galloping at higher wind speeds;
- wind–rain vibrations at lower wind speeds.

An important dimension-free parameter in aerodynamics is the Scruton figure, which is defined by

$$Sc = \frac{m\xi}{\rho d^2} \tag{5.25}$$

where
m = cable mass per unit
ξ = critical damping of the cable, determined by the *AVM*
ρ = air density ($1.25\,kg/m^3$)
d = cable diameter

High values of the Scruton figure are an indication of the fact that the oscillations are suppressed or the beginning of instability is only reached for higher wind speeds. As apparent from equation (5.25), the damping is the decisive factor for the Scruton figure. Values for damping of $<0.3\%$ were measured for very long cables. Many vibration problems on cables can be attributed to damping that is too low.

The American Post Tensioning Institute (PTI) recommends the calculation of a critical wind speed for the assessment of the cables with regard to their susceptibility to vibrations:

$$V_{\text{crit}} = kf_n d\left(\frac{m\xi}{\rho d^2}\right) \tag{5.26}$$

If the actual eigenfrequency of a cable or tendon and the respective damping are known by means of a measurement, this can be checked with regard to susceptibility to vibrations. For the Scruton figure values >10 are regarded as safe. Vibration problems of cables can be counteracted by changing the eigenfrequency or the damping value. For example, the damping values can be increased by installation of connecting wires or damping elements.

5.6 DAMAGE IDENTIFICATION AND LOCALIZATION

Recently there has been significant attention on structural health monitoring (SHM) of engineered systems, especially focused on the space/aerospace, mechanical and automotive systems [15–17]. Members of the civil engineering community participated in these initiatives with limited success. It is intended to take advantage of the synergy in cross-disciplinary research on condition and damage assessment, but there is a need to recognize any distinctions between 'constructed' or 'manufactured' engineering systems. This is particular in terms of:

- size;
- costs;
- life cycle;
- variability in material properties;
- uncertainties in identification of the system;
- uncertainties in identification of operating and loading environments.

This makes a civil engineering task of damage detection rather complex compared to other sectors.

5.6.1 *MOTIVATION FOR SHM*

Despite the distinction between manufactured and constructed systems, a generic framework is expected for sharing many technologies and algorithms serving health monitoring of all engineered systems. Civil engineers are especially aware of the limitations of their current practice for condition assessment based on visual inspections. Typical routine applications of condition assessment are carried out on bridges, dams, industries or buildings for evaluating seismic vulnerability or post-earthquake damage. The engineering community has long been aware of the limitations and shortcomings of visual inspection in the light of the needs of society. A recent investigation carried out by the Federal Highway Administration (FHWA) in the United States showed a dramatic situation with a striking lack of reliability [18]. Out of the 600 000 bridges in the US network, 160 000 are considered to be deficient. One billion users pass over deficient bridges every day and 10 000 bridges have to be replaced annually at costs of over US$7 billion. In Europe, comprehensive figures do not exist, but by deriving overall figures from national statistics it has been anticipated that a burden of about €7 billion annually rests on European communities from this vital bottleneck of the transportation infrastructure. The US reaction to this situation was the implementation of an NSF–FHWA joint engineering research initiative on advancing states of the art and practices of engineering and management of the highway transportation infrastructure. It is intended to raise a total of US$100 million for a 20 year project on research, particularly focused on the following subjects:

- a long-term bridge performance programme including a representative sample of thousands of bridges;

- instrumentation and permanent monitoring of hundreds of these bridges in conjunction with standardized detail inspection for comparison;
- a long-term programme with a target of 20 years for the permanent collection of performance data of bridges;
- autopsy of decommissioned bridges, where some of the 10 000 bridges replaced would be used for damage tests;
- implementation of a huge resultant database on the monitoring data;
- developing the bridge of the future, which would be less costly and longer lasting;
- development of the next generation decision support and management systems;
- consideration of the following new and important topics, such as global warming, assessment of cables and tendons, terrorist attack and emerging computing technologies.

The target is to create highways for life and to encourage bridge owners to take greater risks based on better decision-making tools in order to reduce the annual investment costs of US$7 billion per year considerably. The US NSF (National Science Foundation) intends to launch an initiative on infrastructure, bridges and components. This is in recognition of this huge societal problem. Further possibilities for health monitoring are included in the NEES project (refer to www.nsf.com/nees) and the successful NSF program on sensors and sensor networks (NSF 03–512), which provides considerable funding in these research areas.

A very similar situation is experienced in Japan, where after the construction boom in the 1980s and 1990s the infrastructure is rapidly ageing. Programmes for structural health monitoring of the infrastructure have recently been launched, complementing the many big singular projects in this subject. The Japanese have largely concentrated on system identification and damage detection. Currently a drive for harmonization and combination of the singular initiative to a major programme can be observed. A first major workshop held at the University of Tokyo in November 2003 can be taken as the start of a major programme that is also looking for international research collaboration with the United States. A special Asia pacific programme has been launched on the subject. It is useful to include European initiatives, which is wanted by all other parties, into this global programme on structural health monitoring of the transportation infrastructure.

5.6.2 CURRENT PRACTICE

Inspections do find signs of damage, such as cracks, spalls, chemical deterioration and corrosion when these become visible and represent the current practice in structural management. However, the relation between such visible signs of damage and the corresponding condition or reliability of the structure is often very difficult to establish. There might be a dramatic difference in the meeting of a certain visual damage for a steel, pre-stressed, or an ordinary reinforced concrete bridge, and often the effect may be observed but the decision making has to be carried out based on heuristics and experience. Its cause may not be identified definitely. Engineers need to know the actual cause of damage or distress and its impact on structural reliability in order to make meaningful management decisions. Most importantly, discovery of deterioration before or at its onset is needed for cost-effective management of structures [19]. There is a lack of correlation between visual appearance and structural reliability for safety. In many cases, this leads to a great shortcoming in assessing the condition based on just a visual inspection. There is a lack of clear and accurate failure scenarios that are able to put the special attention into fatigue prone and fracture critical details. Current guidelines are too broad for reliability reasons and actually lead to uneconomic procedures.

Although considerable research on damage and condition assessment has been conducted, there are still fundamental issues to be resolved. Most research has been in the area of manufactured systems, and the distinctions between manufactured and constructed systems may not be well understood by researchers from mechanical and aerospace fields. Consensus definitions, measures and indices for performance, condition, damage and health over the life cycle of common types of constructed facilities and complete infrastructure systems are necessary for a reliable condition assessment. Also, drift, displacement, crack width and stress levels have typically been used for defining the onset of service ability and damage ability limit states for common constructed facilities. Proven relationships between such measurable indices and actual facility performance have not been established. There is evidence that actual deflection, drifts and stress at a constructed facility are very different from what would have been predicted in design. Both the actual demands and the capacities of a constructed system are often different from the estimates on which the design would have been based (refer also to the five European FP projects SIMCES, IMAC and SAMCO).

A review of damage indices that have been proposed for nuclear, aerospace, mechanical, offshore and civil engineering structures provides an excellent overview [17,20–22]. The sensitivity of vibration-based indices to various levels of damage has been evaluated by DeRoeck (bridge Z24 in Switzerland) (see section 5,3,2), Wenzel *et al.* (Regau Bridge in Austria) [23] and the Los Alamos Laboratories (on a steel girder bridge in New Mexico) [20]. The changes in modal strain energy caused by damage, evaluated by processing the modal data, are utilized. More recently research is concentrating on the data of permanent monitoring systems established in a number of instrumented bridges like the Seto Bridges in Japan, the Ting Kao Bridge in Hong Kong and the Europa Bridge and the St Marx Viaduct in Austria. It has been observed that the damage detection initiatives in Asia and the United States are research driven, whereas the European initiatives so far are based on end user interest.

5.6.3 CONDITION AND DAMAGE INDICES

Displacement and strain influence coefficients are conceptual, physics-based indices representing structural characteristics that are meaningful to engineers, and they are often analytically predicted for designing or evaluating structures. In addition, monitoring the redistribution of intrinsic strains at critical structural systems, members or components, such as movement systems, hangers, maximum response locations, or at boundary and continuity locations, provides valuable information. Intrinsic strains, as well as temperatures, at these critical components, which have a direct impact on the structural reliability, should be monitored for a sufficiently long duration so that signs of damage and deterioration may definitely be identified.

The same indices may actually be measured by various experimental techniques, intermittently as well as continuously. Changes in influence coefficients of an assembly of critical responses provide a powerful vehicle for a measure of the objective of global condition and health as long as they are experimentally reliably determined.

Many engineers consider damage as changes in the effective material properties within a structure, and many non-destructive technologies that can successfully characterize the *in situ* properties of construction materials, even through covers and other obstructions, have been developed. Such advances, which are generally utilized for localized condition evaluations and indices, are very useful when used in the context of detecting the onset of deterioration, such as the initiation of a corrosive environment in a reinforced concrete element or the beginning of deterioration of a chemical bond between steel and concrete in composite bridges. However, if the critical deterioration mechanism or the critical regions and responses of recurring constructed facilities that need to be monitored have not yet been established, local scans and measurements cannot feasibly and effectively address the problem of the global condition and damage assessment. Even if it were possible to conduct a complete local scan throughout a constructed facility, for effective management it is also necessary to understand how local damage affects the complete system performance. Global damage may be described as phenomena that distinctly and irreversibly influence the force displacement responses of the critical regions of a structure, signifying the onset of the safety limit states. A comprehensive discussion of damage indices can be found in the literature (refer to references [16] and [24]).

Nonlinear material damage indices based on a continuum approach [25] and indices based on post-yield displacements or hysteretic energy dissipation by structural elements have been proposed for assessing bridges subjected to ambient vibrations [22,26]. More recently, numerous indices based on non-parametric characterizations such as those based on neural networks have been proposed [27]. These indices require the analysis of recorded response time histories during a damaging event, and whether they can effectively serve for the assessment of obscure damages following an earthquake needs to be verified. While no soil (foundation) structure interaction system can be strictly linear, the justification for using linearized indices is that most highly redundant constructed facilities behave linearly in the global sense shakedown, even when many local nonlinearities, such as due to localized damage, may exist. Recent experience from the European project IMAC (Project GRD1-2000-25654) showed that the simplified linear approach taken by US researchers might have a limited value, particularly for pre-stressed concrete beams, which represent a major area of interest in Europe.

For the general infrastructure management problem, linearized indices that are possible to conceptualize physically with respect to a shakedown state would be desirable. They may be directly measured or easily extracted from measurements during controlled tests at any time, and easily correlated to structural performance at the serviceability limit states.

There is a complicated relationship between the life cycle of a facility being monitored, experimental constraints and suitable indices. Meaningful conditions or damage assessment would have to recognize the life cycle stage, and functional and operational parameters of the facility as well as the infrastructure systems that the facility support, any occurrence of accidents, overloads or disasters. Ideally, intermittent applications such as geometry measurements, diagnostic tests and non-destructive evaluation applications need to be integrated with continuous life cycle health monitoring for condition and damage assessment of critical facilities.

It is necessary to integrate a spectrum of experiments and indices, and to monitor a facility over a long period, preferably starting from construction, for reliable condition and damage assessment. Before completely understanding the structural, foundation and soil systems, the load path, critical load-carrying capacities and possible failure mechanisms, expecting to visit a constructed facility at some stage of its life cycle and evaluating its condition by conducting a single experiment to measure or compute a single index is not realistic. A well-coordinated and well-structured integration of experiment, analysis and information technologies in the context of structural identification becomes critical. It is assessed that the demand for research and development is the highest particular in the field of decision support systems, where the various indices available are rated, sorted and assessed. Neural network applications are the most promising approaches. These are currently in the mature stages of implementation and attract the interest of the research community. The vision is to provide a first integrated system based on neural networks on the market by 2010.

5.6.4 BASIC PHILOSOPHY OF SHM

Current practice in the Los Alamos Laboratories demonstrated by Mr C. Farrar provides in a basic framework that SHM should consider:

- Sensors should not be the limiting factors. Enough sensors with redundancy should be used to cover the eventual loss of information.
- Data interrogation is a key function in SHM. There are no sufficient routines to check and assess the data on quality and consistency.
- Predictive modelling takes its role beside ordinary static or dynamic modelling. The calculation models should be integrated into the process. It is not sufficient any more to have one model only.
- Calculation capacity is no longer the limiting factor. In Los Alamos they are able to solve any matrix of 10 million DOFs within 2 min. There is an extensive co-operation with the University of California San Diego (UCSD) on supercomputing. They have a computing centre furnished by Compaq which is the size of a football field, with world record performance data.
- The focus of future research will be on integrated solutions looking at SHM, from monitoring to decision making.
- New sensors such as MEMS and NEMS play a major role, but they are not yet sufficiently developed.

Current development work concentrates mainly on this research field. Numerous universities and some European research projects have focused on this topic. It is currently assumed that damage can be read in the following parameters:

- changes in fundamental eigenfrequencies in the case of global damage;
- changes in higher eigenfrequencies in the case of local damage;
- change in the damping characteristic in the case of material fatigue;
- changes in modal flexions due to local damage;
- failure of individual frequencies due to transposition;
- change in the mode shape of individual eigenfrequencies.

In the AVM modal analysis and damping analysis are the methods currently most used. In modal analysis the eigenfrequencies are observed over a certain period of time and their changes are graphically represented. The representation is done in so-called trend cards, where measurements are chronologically applied. Unchanged eigenfrequencies are documented by straight (vertical) lines (Figure 5.31).

In order to be able to verify the developments of BRIMOS®, five bridges were artificially damaged in 2001 and the effects on the dynamic characteristic assessed. The damages were mainly caused by severing individual tendons or reinforcing bars. These damages caused a change in the load-bearing system, which is determined by the measurements. In most cases the basic frequencies

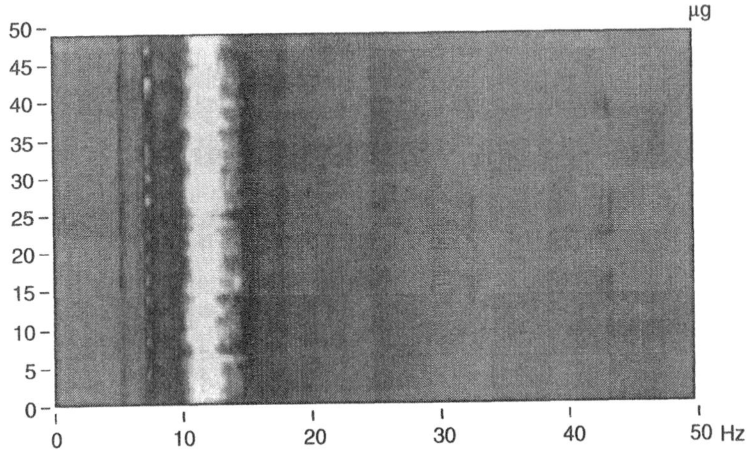

Figure 5.31 Representation of a trend card with an unchanged structural condition

are not concerned; at higher frequencies, however, with their corresponding short-wave vibration forms, distinct changes are noticeable (Figure 5.32). In particular, force transpositions of individual structural members are discernible by means of frequency measurements (trend representation). When an element breaks down, its eigenfrequency drastically drops and the frequency of another element rises due to the increase in the force.

An essential aim of the BRIMOS® recorder development is the establishment of trend cards for numerous structures. The recording of a corresponding data

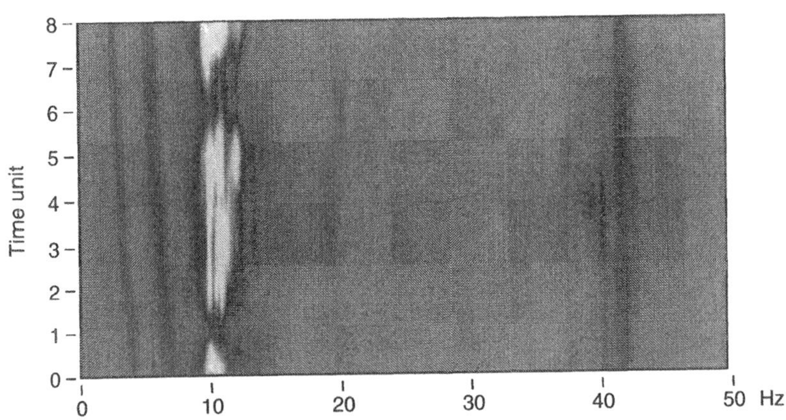

Figure 5.32 Trend card of a progressing damage event

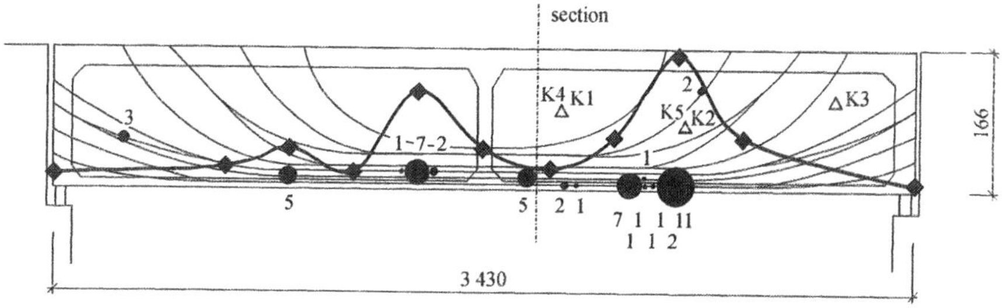

Figure 5.33 Damage localization at the Al flyover at Regau

basis is, however, very time-consuming. The transfer of the experience acquired from damage tests in 2001 is still problematic because the latter were always carried out within one day. The effects of varying environmental influences (temperature) could therefore not be appropriately considered for the interpretation of long-term observations, but these influences are decisive. Furthermore, system-dependent changes (roadworks) have to be considered for the interpretation of long-term records. An accurate documentation of the history of the structure is therefore required.

A further important criterion of damage identification and localization is the interpretation of the damping parameters. In this case it does not proceed from classical modal damping but local damping (spreading of energy) is analysed. This means that the behaviour of every range of the structure should be considered locally. The damping measured is a measure of how the structure deals with the applied energy. If there are cracks in the structural member, energy is converted and higher damping values are noticeable. This phenomenon is, however, not only limited to cracks but also reliably demonstrates other damages, e.g. fractions in pre-stressing steel. Thus it is also possible to assess pre-stressed bridges with uncracked concrete. An example is the bridge over the Western motorway near Regau (S123a), where the damage was determined by measurements taken at the structure. These defects were verified when the structure was subsequently demolished (Figure 5.33).

5.7 DAMAGE PROGNOSIS

Damage diagnosis or structural health monitoring is the process of assessing, the current structural condition of a system. This process aims to continuously monitor the structure's condition, indicating the need for maintenance or change in system operation when, for instance, damage-sensitive features extracted from the monitored system reach a critical level. This process will not provide information, however, about how long a system can continue functioning without failure.

Damage prognosis is a process for predicting the remaining useful life given the measurement and assessment of a system's current structural condition, as well as information concerning the effects of the anticipated future loading environment on system performance. All damage prognosis solutions require the integration of three general technology areas:

1. Predictive modelling
2. Sensing and processing hardware
3. Data interrogation algorithms

The current state-of-the-art tools available in these three categories were not necessarily developed with damage prognosis in mind. Therefore these various technologies must be extended beyond the current state in a synergistic manner with the focus on the damage prognosis goal.

5.7.1 SENSING DEVELOPMENTS

There is a need to develop multiscale active sensing systems that can both detect damage at its onset and evaluate the influence of this damage on the system level response. Here the term 'active' refers to the concept that the sensing system will also provide an input to the structure that will allow a better damage interrogation than provided by the ambient system response. The system must be fault tolerant and non-inclusive. Finally, there must be a methodology for defining the sensing system properties (i.e. bandwidth, resolution, sampling, etc.).

Sensing systems play a major role in all aspects of the damage prognosis effort:

- First, the numerical model constructed for physics-based knowledge needs to be validated using experimental data measured at an actual structure.
- Second, the current damage state of the structure will also be estimated based on the data measured using the sensing system.
- Finally, prediction of the system's future performance requires monitoring and extrapolation of past leading histories for estimating future loading characteristics.

5.7.2 DATA INTERROGATION PROCEDURE FOR DAMAGE PROGNOSIS

Data interrogation includes many aspects of data analysis, signal processing, machine learning techniques, data management and feature extraction. Data interrogation supports final decision making with regard to the system's damage state and remaining life by extracting relevant information both from experimental data measured from sensors and from simulation data generated from numerical models.

5.7.3 PREDICTIVE MODELLING OF DAMAGE EVOLUTION

The issues for predictive modelling are somewhat similar to that of sensing. There is a need to capture the system response on widely varying length and time scales. Also, the fidelity of the predictive models must be increased. Procedures must be developed to condense large-scale models so that the reduced order models can be run on microprocessors directly integrated with the sensing systems.

Proof of concept experiments should be conducted at all stages of the development work. Various applications will require different approaches in the experiments. Nevertheless, cross-fertilization through industrial sectors should be expected.

Uncertainty quantifications are another important aspect of damage prognosis, model validation and reliability analysis. To predict the progression of damage during fatigue loading, a relationship must be established between the system parameters, system responses and damage indicators. Because the costs of developing an empirical relation based on destructive testing of numerous samples are prohibitive, numerical modelling must be employed. This requirement leads to the need for propagation of uncertainty through the models, calibration of the models with respect to physical experiments, assessment of their predictive accuracy and application of reliability analysis for decision making.

Reliability analysis requires the incorporation of a failure model into the final element model to build a relationship between the system response and damage level. The features of this damage model also have uncertainty and are therefore treated probabilistically. The analysis begins with identifying the failure modes and the random variables that contribute to these failures.

To date, the only successful damage prognosis solutions are referred to as empirical damage prognosis where the system of interest has been monitored to failure. Subsequently, regression models are developed that can relate some measured operational parameter to a feature extracted from the measured data relating to the system condition. This approach has been used successfully for rotating machinery applications. The goal of current research is to extend the concept of damage prognosis to a more general class of structures. The first step in this process is to complete a damage prognosis study on a relatively simple structure (post-tensioned concrete beam) and quantify the uncertainty for all steps in the process. It is anticipated that examples from other sectors will cross-fertilize this approach.

5.8 ANIMATION AND THE MODAL ASSURANCE CRITERION (MAC)

5.8.1 REPRESENTATION OF THE CALCULATED MODE SHAPES

The geometry of the structure is taken from a framework programme, where corresponding files are delivered. The transmission of the mode shapes is done in the same way as with the geometric properties; animation of the mode shapes is also possible.

The coordinates of the check points are delivered in the same system of coordinates as used in the framework programme. This results in a network model for the points where the values for the movement in every single form are delivered. These values are fixed values and are only animated via the frequencies (analogous to calculation). Only the individual points that are connected in a linear way are moved to make the comparison between calculation and measurement possible.

5.8.2 GENERAL REQUIREMENTS

Important for the animation of the measured values is a realistic representation and therefore values are not averaged. The animation can run automatically or have the option to stop and continue in steps. Calculation and measurement move synchronously to ensure comparability.

The data are delivered in the following steps:

- a file from the programme FE that contains the geometric data;
- files that contain the information on the mode shapes;
- file with coordinates of the check points;
- file with the values for the individual check points for the eigenfrequencies to be animated;
- files with chronological histories for the check points.

5.8.3 CORRELATION OF MEASUREMENT AND CALCULATION (MAC)

During a comparison of the measured and calculated eigenfrequencies the mode shapes are neglected and therefore cannot be accurately assessed if the measured vibration modes correspond to the calculated ones. Calculation of the MAC offers the possibility to compare the mode shapes of two systems accurately and additionally to localize deviations. The MAC value is the square of the cosine of the angle between two eigenvectors (at one point). The closer this value is to one, the smaller is the angle between the two eigenvectors. In the case where the value is zero they are orthogonal vectors:

Table 5.3 COMAC matrix

	$\text{Mode}_{1R,1M}$	$\text{Mode}_{1R,2M}$	$\text{Mode}_{1R,3M}$	$\text{Mode}_{1R,1M}$	
Node 1	MAC $(1,1)_1$	MAC $(1,2)_1$	MAC $(1,3)_1$	MAC $(n,m)_1$	COMAC(1)
Node 2	MAC $(1,1)_2$	MAC $(1,2)_2$	MAC $(1,3)_2$	MAC $(n,m)_2$	COMAC(2)
Node 3	MAC $(1,1)_3$	MAC $(1,2)_3$	MAC $(1,3)_3$	MAC $(n,m)_3$	COMAC(3)
Node 4	MAC $(1,1)_4$	MAC $(1,2)_4$	MAC $(1,3)_4$	MAC $(n,m)_4$	COMAC(4)
Node n	MAC $(1,1)_n$	MAC $(1,2)_n$	MAC $(1,3)_n$	MAC $(n,m)_n$	COMAC(n)
	MAC (1,1)	MAC (1,2)	MAC (1,3)	MAC (nR,mM)	

$$\text{MAC}(n,m)_n = \frac{\left|(\psi_{\text{R}n})_n^{\text{T}}\ (\psi_{\text{M}m})_n\right|^2}{(\psi_{\text{R}n})_n^{\text{T}}\ (\psi_{\text{R}n})_n\ (\psi_{\text{M}m})_n^{\text{T}}\ (\psi_{\text{M}m})_n^{\text{T}}} \tag{5.27}$$

The individual MAC values are arranged in the so-called COMAC matrix (Table 5.3), with R standing for calculation and M for measurement.

In order to determine the correct COMAC factor those columns of the matrix that result in a low MAC value must be sorted out. Before results of a modal analysis can be compared and correlated to those of an FE calculation, two basic incompatibilities between the resulting data records have to be cleared up:

- varying numbers of eigenvectors;
- complex eigenvectors (measurement)–real eigenvectors (calculation).

5.8.4 *VARYING NUMBER OF EIGENVECTORS*

Usually real structures are systems with many more degrees of freedom than shown in the calculation. It is possible to either adapt the measured vectors to the system size or reduce the calculated eigenvectors to the dimension of the measurement. Only after this correlation can further considerations and comparisons between measurement and calculation be done.

During reduction to the size of the measurement every check point is first of all allocated to a node of the FE model that is geometrically closest. The node with the smallest distance from the check point is chosen. The data of the FE analysis are thus reduced to the size of the measurement model and a reduced data record of the FE calculation is generated.

5.8.5 *COMPLEX EIGENVECTOR MEASUREMENT*

During modal analysis of real structures complex eigenvectors occur due to existing damping effects and nonlinearities. These complex mode shapes cannot

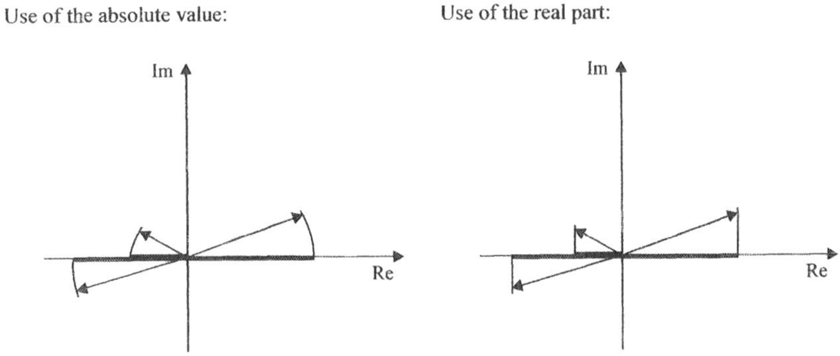

Figure 5.34 Representation of a complex number level

be easily compared to the mode shapes of FE calculation because here real mode shapes are calculated.

In contrast to the real eigenvectors that describe the natural vibrations of a structure in the form of maximum displacements at every node, complex mode shapes additionally supply a phase position for every displacement. The individual nodes of the structure do not oscillate in phase (or in antiphase) to each other, but every node has its own phase position. Therefore it is not easily possible to draw a displacement picture of a complex eigenvector as the individual node displacements do not reach their maximum at the same time.

Before these complex natural vibration forms can be compared and correlated with the real eigenvectors from an FE calculation, they have first to be 'realized' or made real. An existing phase displacement must be blanked out.

The components of real eigenvectors have a phase position of either 0 or 180°; i.e. the respective indexes would be situated on the real axis in a representation of the complex number level. One of the simplest possibilities to convert complex eigenvectors into real ones is to consider only the real part or the absolute value of the measured eigenvector (Figure 5.34).

If individual components of the eigenvector have great phase displacements, i.e. they enclose a big angle with the real axis, such a reduction is not admissible. The evaluation of such measured data is a problem that can only be coped with by individual considerations.

5.8.6 SELECTION OF SUITABLE CHECK POINTS USING THE MAC

The selection of possible installation points is limited by the number of available sensors as well as the accessibility of the check points. The aim is to select those installation points where the measured values enable a clear

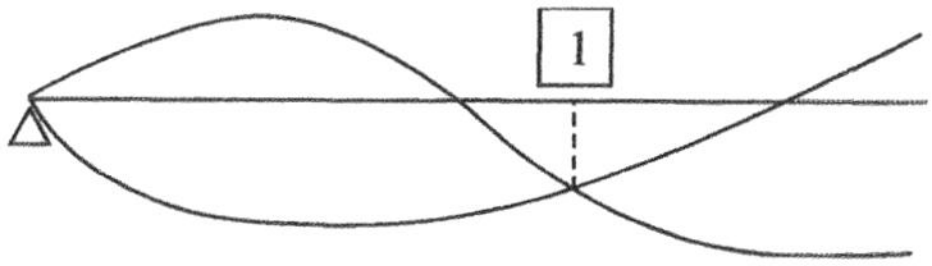

Figure 5.35 Badly placed check point

distinction to be made of the various mode shapes. Therefore those points have to be sorted out where the same or similar structural responses are measured for different mode shapes, as shown in Figure 5.35. This is problematic above all when mode shapes with similar eigenfrequencies are concerned; this effect is called 'aliasing'.

Such critical points can be separated out by means of an FE analysis. The MAC matrix provides information on the orthogonality of the observed eigenvectors (Tables 5.4 and 5.5). The value 1 at the position $[I, j]$ in the MAC matrix means that the vector I is identical to vector j regardless of its absolute value. Zero values, however, identify orthogonal vectors. Node 2 is not suitable as a measuring location because according to the MAC matrix the first natural mode can only be distinguished from the third natural mode with difficulty.

Table 5.4 MAC matrix: comparison of calculation and measurement

Node number	$Mode_{1R}$	$Mode_{2R}$	$Mode_{3R}$	$Mode_{nR}$
$Mode_{1R}$	MAC (1,1)	MAC (1,2)	MAC (1,3)	MAC (1,n)
$Mode_{2R}$	MAC (2,1)	MAC (2,2)	MAC (2,3)	MAC (2,n)
$Mode_{3R}$	MAC (3,1)	MAC (3,2)	MAC (3,3)	MAC (3,n)
$Mode_{nR}$	MAC (n,1)	MAC (n,2)	MAC (n,3)	MAC (n,n)

Table 5.5 Ideal condition of the standard matrix

Node 1	$Mode_{1R}$	$Mode_{2R}$	$Mode_{3R}$	$Mode_{nR}$
$Mode_{1R}$	1	0	0	0
$Mode_{2R}$	0	1	0	0
$Mode_{3R}$	0	0	1	0
$Mode_{nR}$	0	0	0	1

Example:

Node 2	$Mode_{1R}$	$Mode_{2R}$	$Mode_{3R}$	$Mode_{nR}$
$Mode_{1R}$	1	0.15	0.93	0.20
$Mode_{2R}$	0.15	1	0.32	0.15
$Mode_{3R}$	0.93	0.32	1	0.24
$Mode_{nR}$	0.20	0.15	0.24	1

If the mode shapes are arranged in the MAC matrix according to the size of their eigenfrequency (the first mode shape has the lowest eigenfrequency), those values further apart from the diagonal can be disregarded. It is advisable to calculate the differences between the individual eigenfrequencies and to mark the points with the lowest difference in the MAC matrix. By this procedure those problematic points of the MAC matrix can be identified that are combinations with a low-frequency difference and a small solid angle. Points showing these unfavourable properties can be removed.

In this connection it has to be mentioned that the planning of measurement installations is best carried out on the basis of the experience of the check personnel. In the last few years the best results were achieved using this procedure.

5.9 AMBIENT VIBRATION DERIVATIVES (AVD®)

AVD stands for 'ambient vibration derivative' and was developed by the authors in the scope of the research work for BRIMOS®. It is modelled on the aerodynamic derivatives method used in wind engineering. Conclusions on possible instabilities are drawn from changed system parameters, based on changed test configurations. In the wind tunnel this is obtained by an increase in load, i.e. a rise in the wind speed. For buildings it is possible to reduce stiffness instead of increasing the load.

5.9.1 AERODYNAMIC DERIVATIVES

The aerodynamic derivatives method was proposed by Scanlan and Simiu [28] and has been accepted worldwide. System parameters are extracted from time series acquired by wind-tunnel tests by means of statistical methods, among others the random decrement technique (RDT), and applied in a chart comparing the change in load. From this a trend is discernible, which enables conclusions to be made on the behaviour of the structure. What is especially important are the factors that describe the change in damping with increasing wind load. If the damping tends towards zero, instability exists.

5.9.2 APPLICATIONS OF THE AVM

For the AVM two possible applications are conceivable, which are, on the one hand, the load-dependent change of system parameters (load increase and dynamic stresses) and, on the other, the time-dependent change of parameters (changes of stiffness by damages):

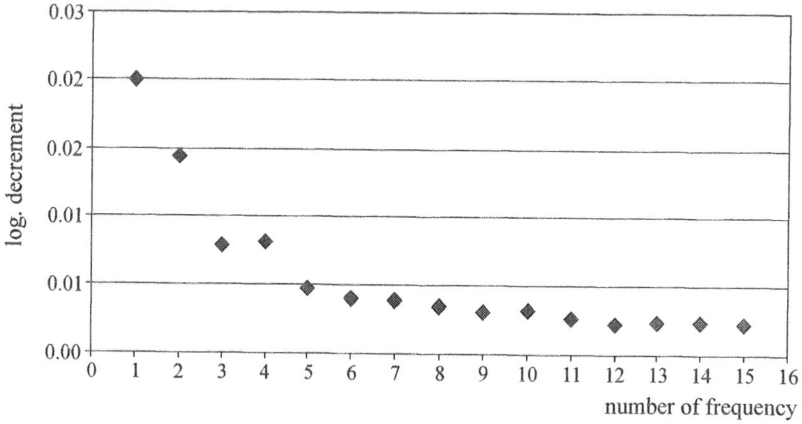

Figure 5.36 Damping versus eigenfrequency of a cable

- *Load changes.* For load changes the procedure could be analogous to the wind-tunnel. The requirements of the saved signal are to be determined by tests.
- *System dependence.* In contrast to load dependence no constant system can be assumed. If white noise is assumed as the input quantity, changes in the signal can be attributed to system-dependent changes. The representation could be analogous to the aerodynamic derivatives, e.g. the change of damping over time (Figure 5.36). What would certainly be interesting is the representation of the changes in acceleration due to the change in the system (deformations). Further parameter analyses are still to be considered in this connection.

5.9.3 *PRACTICAL IMPLEMENTATION*

If there is a possibility of using test loads, the procedure can be analogous to wind-tunnel tests. Usually it will be necessary to go back to measurements. Temperature compensation has to be carried out to calibrate dependent parameters.

The process requires a large amount of calculation and is also very time-consuming with regard to testing. The sample time has to be increased considerably.

This process is still under development with the main focus on:

- frequency changes over time in the frequency domain (the result is trend cards, see Figures 5.31 and 5.32);

- change in effort over time in the time domain (the result is a shift in the system axis);
- load-dependent frequency modulation;
- damping ratios of subsequent frequencies;
- intensity diagrams.

REFERENCES

1. Cooley, J.M. and Tukey, J.W. (1965) An algorithm of the machine calculation of complex Fourier Series. *Mathematics of Computation*, **19**(90), 297–301.
2. Peeters, B. (2000) *System Identification and Damage Detection in Civil Engineering*. PhD thesis, Katholieke Universiteit Leuven, Leuven, Belgium.
3. Peeters, B. and De Roeck, G. (2000) Reference based stochastic subspace identification in civil engineering. *Inverse Problems in Engineering*, **8**(1), 47–74.
4. Peeters, B. and De Roeck, G. (1999) Reference-based stochastic subspace identification for output-only modal analysis. *Mechanical Systems and Signal Processing*, **6**(3), 855–78.
5. Maeck, J. and De Roeck, G. (2003) Description of Z24 benchmark. *Mechanical Systems and Signal Processing*, **71**(1), January, 127–31.
6. Maeck, J. (2003) *Damage Assessment of Civil Engineering Structures by Vibration Monitoring*. PhD thesis, Katholieke Universiteit Leuven, Leuven, Belgium.
7. Krämer, C., De Smet, C.A.M. and De Roeck, G. *Z24 Bridge Damage Detection Tests*. Proceedings of IMAC 17: International Modal Analysis Conference, February 1999, Kissimmee, Florida, pp. 1023–9.
8. Teughels, A., Maeck, J. and De Roeck, G. (2002) Damage assessment by FE model updating using damage functions. *Computers and Structures*, **80**(25), October, 1869–79.
9. MATLAB (2000) *Matlab Optimization Toolbox User's Guide*, Version 2.1 (Release 12.1), MathWorks Inc., Natick, Massachusetts, http://www.mathworks.com/products/optimization/.
10. Nocedal, J. and Wright, S.J. (1999) *Numerical Optimization*, Springer, New York.
11. Conn, A.R., Gould, N.I.M. and Toint, P.L. (2000) *Trust-Region Methods*, SIAM Society and Mathematical Programming Society, Philadelphia, Pennsylvania.
12. MATLAB (2001) *The Language of Technical Computing*, Version 6.1 (Release 12.1). MathWorks Inc., Natick, Massachusetts, http://www.mathworks.com/.
13. Fox, R. and Kapoor, M. (1968) Rate of change of eigenvalues and eigenvectors. *American Institute of Aeronautics and Astronautics Journal*, **6**, 2426–9.
14. ANSYS (2003) *Robust Simulation and Analysis Software*, http://www.ansys.com/, Release 6.1, ANSYS Incorporated.
15. Chang, F.-K. (1999) *Structural Health Monitoring: A Summary Report of the First International Workshop on Structural Health Monitoring, September 18–20, 1997*, Structural Health Monitoring 2000, Stanford University, Palo Alto, California.
16. Aktan, A.E., Catbas, F.N., Grimmelsman, K.A. and Tsikos, C.J. (2000) Issues in infrastructure health monitoring for management. *Journal of Engineering Mechanics, ACSE 126*, **7**, 711–24.
17. Chase, S.B. and Aktan, A.E. (2001) *Health Monitoring and Management of Civil Infrastructure Systems*. Proceedings of SPIE Volume 4337, 8/2001.
18. FHWA (2001) Official Website, Bridge Technology, Federal Highway Administration, Washington, DC.

19. Frangopol, D.M. and Das, P.C. (1999) Management of bridge stocks based on future reliability and maintenance costs, in *Bridge Design, Construction, and Maintenance*, Institution of Civil Engineers, Thomas Telford, London, pp. 45–58.
20. Doebling, S.W., Farrar, Ch.R., Prime, M.B. and Shevitz, D.W. (1996) *Damage Identification and Health Monitoring of Structural and Mechanical Systems from Changes in Their Vibration Characteristics*. A Literature Review, Los Alamos National Laboratory Report LA-13070 MS, Los Alamos Laboratories, May 1996.
21. Farrar, C.R., Doebling, S.W. and Prime, M.B. (1998) A summary review of vibration-based damage identification methods. *The Shock and Vibration Digest*, **30**(2).
22. Wenzel, H. (2001) *Damage Detection and Assessment by Ambient Vibration Monitoring*. OECD–NEA Workshop on the Seismic Re-evaluation of all Nuclear Facilities, Joint Research Centre of the European Union (EC JRC) Ispra, Italy, March 2001.
23. Wenzel, H., Geier, R. and Eichinger, E. (2001) Endbericht: Untersuchungen *anlässlich des Abbruches ausgewählter Tragwerke*, von VCE and TU Wien, Wien, October 2001.
24. Farrar, C., Worden, K. and Sohn H. (2003) *Applications of Nonlinear System Identification to Structural Health Monitoring*. 2nd European Workshop on Structural Health Monitoring, Munich, Germany.
25. Mazars, J. (1986) A description of micro- and macroscale damage on concrete structures. *Engineering Fracture Mechanics*, **25**(5–6), 729–37.
26. DeRoeck, G. (1998) *SIMCES – System Identification to Monitor Civil Engineering Structures*. BRITE/EURAM Project Number BBW 96.0202, 1 January 1997–31 December 1998.
27. Nakamura, K. (1998) Neural processing in the subsecond time range in the temporal cortex. *Neural Computation*, **10**(3), 567–95.
28. Scanlan, R.H. and Simiu, E. (1996) *Winds Effects on Structures: Fundamentals and Applications to Design*, 3rd edition (Wiley-Interscience, New York).

6

Theoretical Bases

Every load-bearing structure not only vibrates due to dynamic superimposed loads but also a 'quasi stationary' structure reacts on excitations from vibrations that are always present in nature. These so-called 'ambient' excitations have the properties of white noise in the statistic average – all relevant frequencies are represented in the response spectrum with almost equal energy content. The minor vibrations a structure shows due to these ambient excitations can be registered by modern highly sensitive acceleration sensors.

Dynamics, i.e. the science of movements under the influence of forces, is often not very familiar to civil engineers, because they normally use statistic considerations for the solution of their tasks. In building design most dynamic problems (earthquakes, wind, waves, etc.) are usually treated by means of static substitute procedures (e.g. multiplication of static equivalent loads with factors). With this procedure the maximum member forces and deformations occurring in a structure due to dynamic influences can be approximately recorded, but the vibration behaviour itself can only be modelled by dynamic analysis.

Inferring a mathematical model from observations and studying their properties is really what science is about. This model may be of more or less formal character, but it has the basic ability of being able to link observations together into some pattern. Thus, the process dealing with the problem of building mathematical models of systems based on observed data from the systems is called 'system identification'. An indispensable part of this process is the transfer function, which represents the system properties linked to specified initial and boundary conditions. This in combination with a given input should approximate the corresponding output as closely as possible. From an engineering point of view the transfer functions form a powerful tool for solving

Ambient Vibration Monitoring H. Wenzel and D. Pichler

inverse problems (system identification and additional damage detection), for simulation, prediction and control of the structural response, for modelling both time-invariant linear and time-varying nonlinear systems, for solving differential equations, etc. As a consequence thereof transfer functions become more and more important in the development of decision support systems.

6.1 GENERAL SURVEY ON THE DYNAMIC CALCULATION METHOD

The response of a structure to dynamic influences is determined by the kind of influence and by the properties of the structure itself. Dynamic influences can occur as variable (operating stress) and as extraordinary events (earthquakes), as defined by ON B 4040. The dynamic properties of the structure can be described by eigenfrequencies, mode shapes and transmission characteristics. They are, on the other hand, determined by stiffness, mass and damping. The response of the structure consists of stresses (member forces) and vibrations (oscillations = temporally variable modifications of form); only the latter, however, can be directly measured (Figure 6.1).

Figure 6.2 gives a general overview of the usual dynamic calculation methods (without claim to completeness). The most important step, which forms the

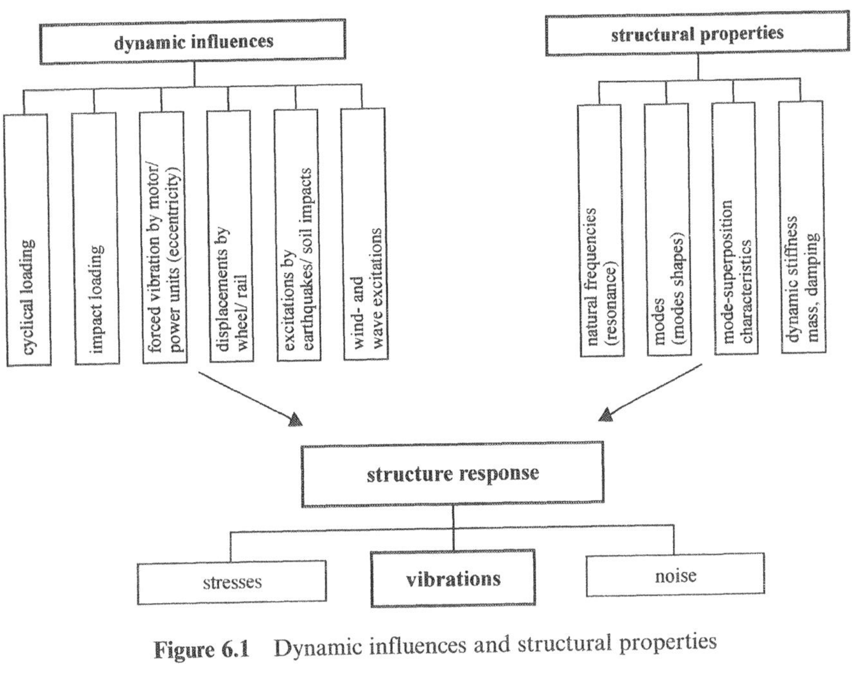

Figure 6.1 Dynamic influences and structural properties

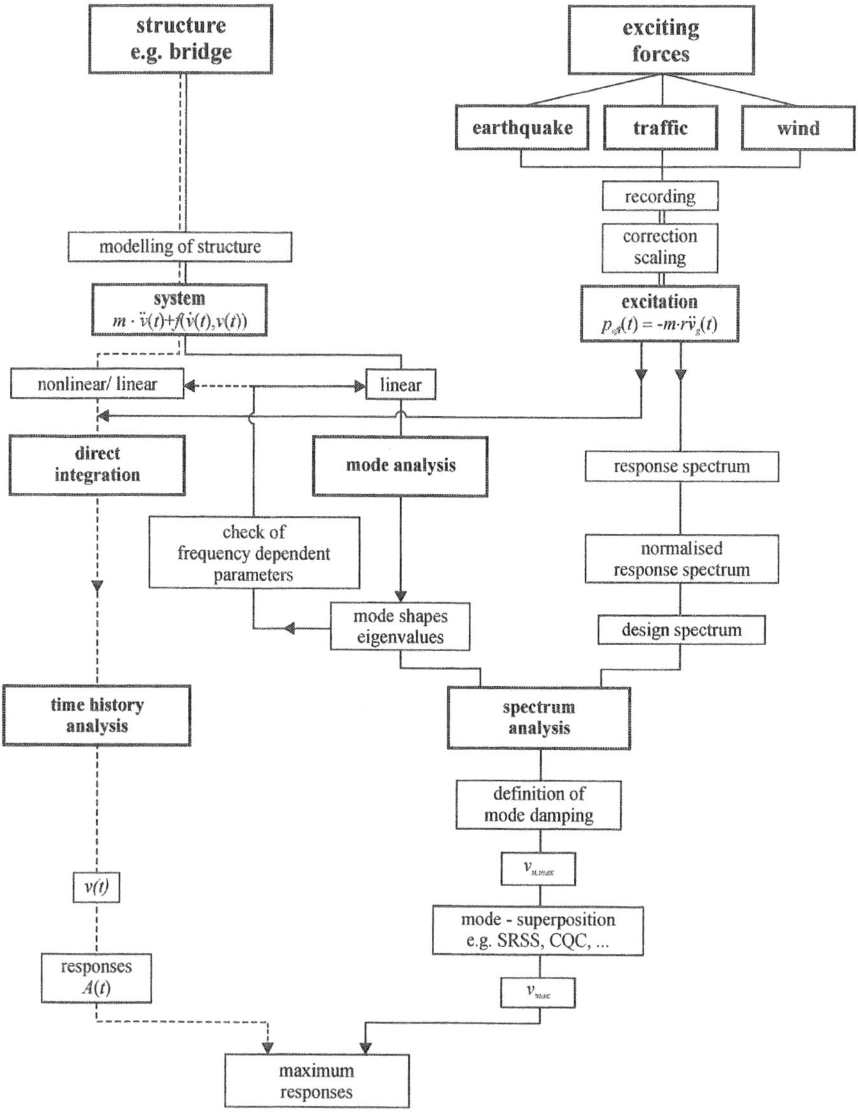

Figure 6.2 Dynamic calculation methods

basis of all calculation methods, is the correct establishment of models. For this purpose the stiffness and masses as well as the bearings of the structures need to be registered sufficiently accurately. It is very difficult to calculate the influence of damping, but empirical values and measuring results can serve as a basis.

$$[m]\cdot[\ddot{x}]+[r]\cdot[\dot{x}]+[c]\cdot[x]=[F(t)]$$

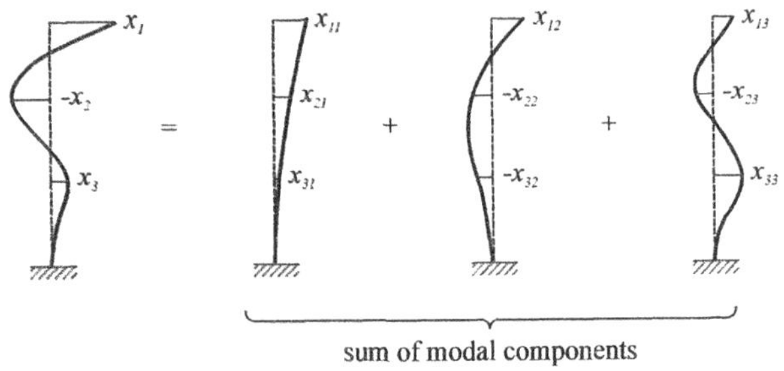

Figure 6.3 Composition of an oscillation from mode shapes

The dynamic calculation methods can basically be divided into two groups: linear and general (linear and nonlinear) methods, where nonlinearities can be caused by the structure (e.g. boundary conditions) or by the material (material laws). As general dynamic calculation methods in the form of nonlinear time-history analyses for the determination of structural responses are very time-consuming, usually dynamic calculations are usually carried out linearly by means of the response spectrum method. Here in a first step the eigenfrequencies and mode shapes are determined by modal analysis. The maximum structural response to an external influence is received by superposition of the mode shapes multiplied by the spectral values of the eigenfrequencies (Figure 6.3).

6.2 SHORT DESCRIPTION OF ANALYTICAL MODAL ANALYSIS

Modal analysis, i.e. the determination of eigenfrequencies and mode shapes (modes) of a structure, can be carried out by means of different methods, as shown in Figure 6.4. A prerequisite for all methods is linear behaviour of the structure. Energetic approaches, as, for example, the Rayleigh method, can be principally applied to SDOF (single DOF) and MDOF (multiple DOF) systems, but an exact solution is generally only possible in the first case. Direct solution of the equation of motion is, however, always possible if the damping matrix is either negligible or represented as a linear combination of the mass and stiffness matrix.

In practice the finite element method (FEM) is universally used today for numerical treatment of beam and shell structures. The FEM is also used for nonlinear problems for calculation of the structural response to general stresses.

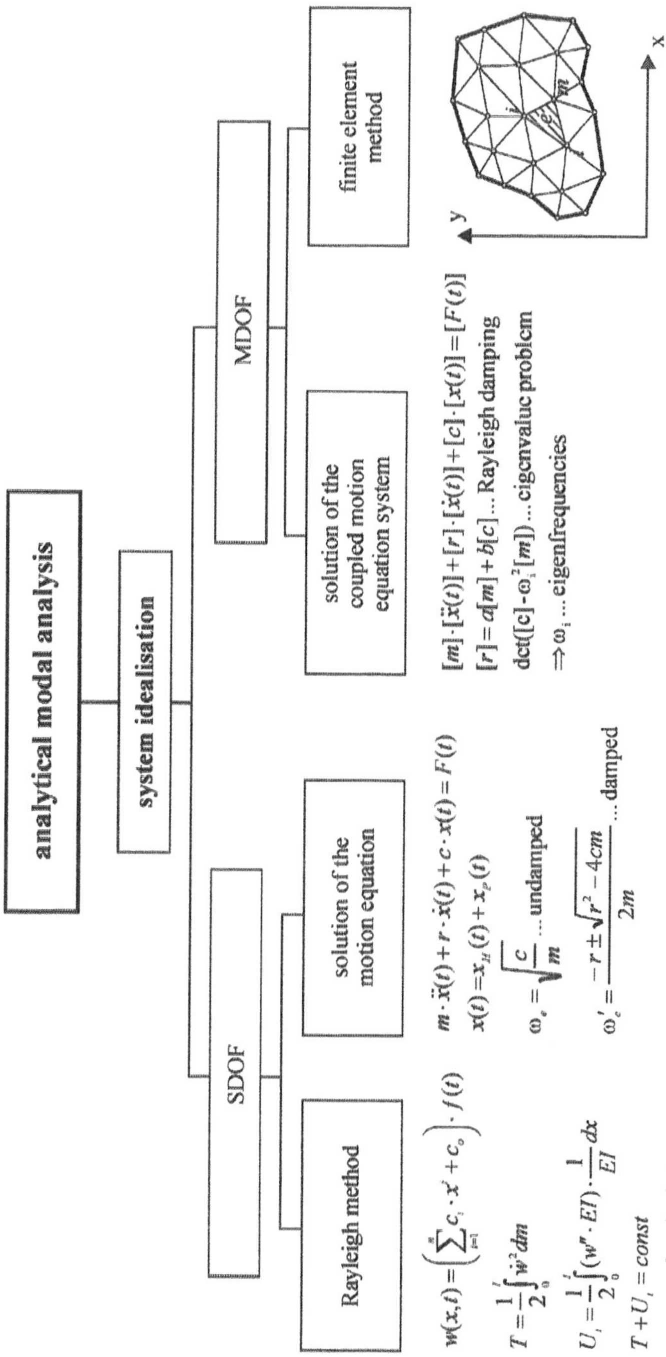

Figure 6.4 Method of analytical modal analysis

6.3 EQUATION OF MOTION OF LINEAR STRUCTURES

As eigenfrequencies and mode shapes are structural properties independent from stress (vibrational signature), it is only natural to use them for the assessment of the maintenance condition of structures. The basis for its determination is the general equation of motion of the structure that is to be examined.

Newton formulated the necessary basic laws in three axioms in the seventeenth century:

1. If all forces affecting a body are in equilibrium, the following applies: $a(t) = 0$; $v(t) = \text{constant}$.
2. The temporal force influence is proportional to the modification of impulse: $F(t)\,\mathrm{D}t = \mathrm{D}[m\,v(t)]$
3. Actio = Reactio

   with
   a = acceleration
   v = velocity
   F = force (influence)
   t = time
   m = mass

The second axiom can be represented in the following formula by transformation and consideration of the critical value $\mathrm{D}t \to 0$: $F(t) = m\,a(t)$. In d'Alembert's style the equilibrium condition can be formulated as $m\,a(t) - F(t) = 0$.

6.3.1 *SDOF SYSTEM*

If d'Alembert's equilibrium consideration is applied to an SDOF system considering a damping r proportional to velocity and a spring constant c, the equation of motion is obtained for a forced damped vibration. The latter has the form of a linear inhomogeneous differential equation of the second order with constant coefficients and can therefore be solved by the simple statement $x(t) = x_h(t) + x_p(t)$ in the case of a harmonic excitation (Figure 6.5).

The solution $x_h(t)$ of Euler's homogeneous differential equation is obtained by an exponential statement $e^{\lambda t}$, which describes the so-called transient effect (flowing back process to the static equilibrium condition). The particular solution is, depending on the form of the disturbance (influence) $F(t)$, a constant or a harmonic function of t corresponding to the disturbance. If $F(t)$ is an arbitrary (non-harmonic) influence, the solution $x(t)$ can generally be represented by a so-called convolution or Duhamel integral.

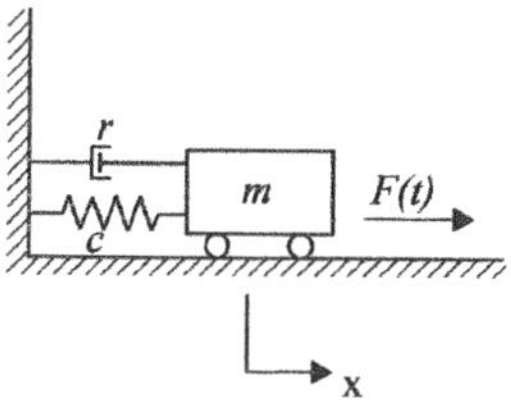

m ... mass
c ... spring stiffness (linear)
r ... damping (velocity proportional)

equilibrium of force at t

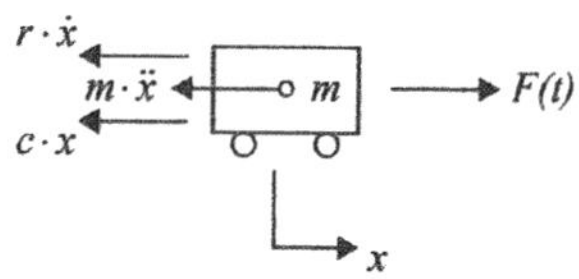

$$\dot{x} = \frac{dx}{dt}$$

$$\ddot{x} = \frac{d^2x}{dt^2}$$

$$\boxed{m \cdot \ddot{x} + r \cdot \dot{x} + c \cdot x = F(t)}$$

equation of motion of a forced damped vibration of a SDOF-System

$\omega = \sqrt{\frac{c}{m}}$ ... natural cyclic frequency (undamped)

$D = \frac{r}{2\sqrt{m \cdot c}}$ ... damping according to Lehr

$$\boxed{\ddot{x} + 2 \cdot D \cdot \omega \cdot \dot{x} + \omega^2 \cdot x = \frac{F(t)}{m}}$$

transformed equation of motion

solution: $x(t) = x_h(t) + x_p(t)$

$x_h(t)$ ...solution of a homogeneous equation
$x_p(t)$ ...particular solution

Figure 6.5 Equation of motion of an SDOF system

6.3.2 MDOF SYSTEM

The equation of motion can be established for a system with MDOFs with an analogous procedure (Figure 6.6). A differential equation system linked via the stiffness matrix is obtained, which can no longer be solved by a simple statement, as in the case of the SDOF system. The mass matrix $[m]$ of the MDOF system meets the criteria of a positively definite diagonal matrix. The stiffness matrix $[c]$ is symmetrical and positively definite according to the Maxwell–Betti theorem.

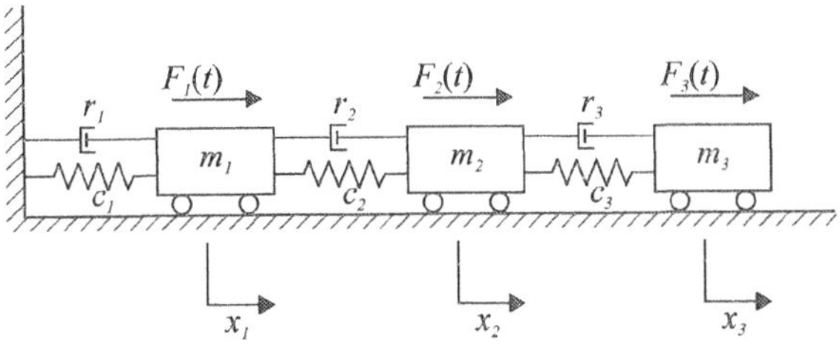

$$[m] = \begin{bmatrix} m_1 & 0 & 0 \\ 0 & m_2 & 0 \\ 0 & 0 & m_3 \end{bmatrix}$$ ... mass matrix

$$[c] = \begin{bmatrix} c_1 + c_2 & -c_2 & 0 \\ -c_2 & c_2 + c_3 & -c_3 \\ 0 & -c_3 & c_3 \end{bmatrix}$$ ... stiffness matrix

$$[r] = a \cdot [m] + b \cdot [c]$$ ... damping matrix (proportional)

$$[x] = \begin{bmatrix} x_1 \\ x_2 \\ x_3 \end{bmatrix} \qquad [\dot{x}] = \begin{bmatrix} \dot{x}_1 \\ \dot{x}_2 \\ \dot{x}_3 \end{bmatrix} \qquad [\ddot{x}] = \begin{bmatrix} \ddot{x}_1 \\ \ddot{x}_2 \\ \ddot{x}_3 \end{bmatrix}$$

$$[F(t)] = \begin{bmatrix} F_1(t) \\ F_2(t) \\ F_3(t) \end{bmatrix}$$

$$\boxed{[m] \cdot [\ddot{x}] + [r] \cdot [\dot{x}] + [c] \cdot [x] = [F(t)]}$$

equation of motion for forced damping vibration of the MDOF-System

Figure 6.6 Equation of motion of an MDOF system

The solution of the equation of motion is carried out in three steps:

1. Decoupling of the differential equation system (return to the SDOF systems).
2. Solution of the decoupled differential equations.
3. Superposition of the individual solutions to the total solution.

The decoupling process is done by determination of the eigenfrequencies and mode shapes. They are obtained by solving the eigenvalue problems $([c] - \omega_i 2 \cdot [m]) \cdot [a_i] = 0$. This homogeneous linear equation has a non-trivial solution only if the denominator determinant of the system det $([c] - \omega_i 2 \cdot [m])$

disappears. For a system with n degrees of freedom (masses), n eigenfrequencies and eigenvectors (mode shapes, modes) have to exist.

The decoupled differential equation system, which is obtained by multiplying the coupled differential equation system by the modal matrix $[a]$ (composed of modal forms) and its transpose $[a]^T$ due to the orthogonality $[a_j]^T \cdot [m] \cdot [a_i] = 0$ and $[a_j]^T \cdot [c] \cdot [a_i] = 0$ for $j \neq i$, can be solved in the same way as n SDOF systems. The total solution of the MDOF system is obtained by superposition of the individual solutions $Y_i(t)$, $i = 1, \ldots, n$ to $[x(t)] = [a] \cdot [Y(t)]$.

6.3.3 INFLUENCE OF DAMPING

In a damped system complete decoupling is only possible if the damping matrix is proportional to the mass and stiffness matrix. This damping form is also called Rayleigh damping.

For the special case $a = 0$, i.e. the damping matrix is only proportional to the stiffness matrix (also called relative damping), higher eigenfrequencies are damped more quickly. In the case of $b = 0$, i.e. with proportionality only to the mass matrix (absolute damping), lower eigenfrequencies are, however, damped more quickly.

The condition $[r] = a \cdot [m] + b \cdot [c]$ is a sufficient but not absolutely necessary criterion for decoupling the equation of motion. In a general case the damping matrix cannot be diagonalized simultaneously with the mass and stiffness matrix. In slightly damped systems, as exist mostly in the building trade, non-diagonal terms can, however, be neglected. As it is usually quite difficult to establish damping of individual eigenfrequencies, a constant modal damping ratio of $\xi_i = \xi$ is usually assumed. Numerous examinations show that the analytic results obtained using this procedure correspond well to the measurements.

6.4 DYNAMIC CALCULATION METHOD FOR THE AVM

As closed solution procedures – even if they are correspondingly simplified – are very time-consuming for real load-bearing structures, the FEM is used for the analytic part of the AVM.

6.5 PRACTICAL EVALUATION OF MEASUREMENTS

6.5.1 EIGENFREQUENCIES

If the recorded accelerations (e.g. by Fourier transformation) are transmitted from the time domain into the frequency domain, a response spectrum is obtained whose energy peaks are near the eigenfrequencies of the structure. On the basis of the discrete Fourier transformation (DFT) Cooley and Tukey [1] developed the fast Fourier transformation (FFT) in the 1960s [2], which makes a quick numerical spectral analysis of the measured accelerations possible. For application of the FFT method the only requirement is that the measured set of data contains $N=2^M$ values, with M as a natural number [2]. This makes it possible to represent all values n and m as a pair of the discrete Fourier transformation from the complex representation of the Fourier series and the Fourier coefficient

$$p(t_m) = \sum_{n=0}^{N-1} P_n \exp\left(2\pi i \frac{nm}{N}\right) \tag{6.1}$$

and

$$P_n = \frac{1}{N} \sum_{m=0}^{N-1} p(t_m) \exp\left(-2\pi i \frac{nm}{N}\right) \tag{6.2}$$

for the period t_m, $m=0, \ldots, N-1$ in a binary form, considerably reducing the calculation time compared with the DFT [3]. This is already considered when acceleration measurements are carried out.

As real load-bearing structures are three-dimensional, the determination of the mode shapes requires the establishment of a measuring grid over the structure to be examined in order to be able to record the vibration behaviour. If the average of all normalized acceleration spectra of the individual check points obtained by the FFT algorithm are determined and the averaged spectrum is squared, a so-called ANPSD spectrum is obtained [4] (Figure 6.7):

$$\text{APSD}(f_k) = \left(\frac{1}{n} \sum_{i=1}^{n} X_i(f_k)\right)^2 \tag{6.3}$$

with $X_i(f_k)$ the Fourier transform for the frequency f_k at check point i. Therefore

$$\text{ANPSD}(f_k) = \frac{\text{APSD}(f_k)}{\text{APSD}_{\max}} \tag{6.4}$$

Using the averaged algorithm those energy peaks are eliminated in the individual spectra that occur owing to short-term disturbances of white noise (e.g. the truck passage over a bridge). A representative average spectrum is obtained where frequently and strongly occurring eigenfrequencies are dominant. As the ANPSD spectrum consists of discrete values, local maximum values of the ordinates and the corresponding abscissa values, the eigenfrequencies can be determined via a simple program loop.

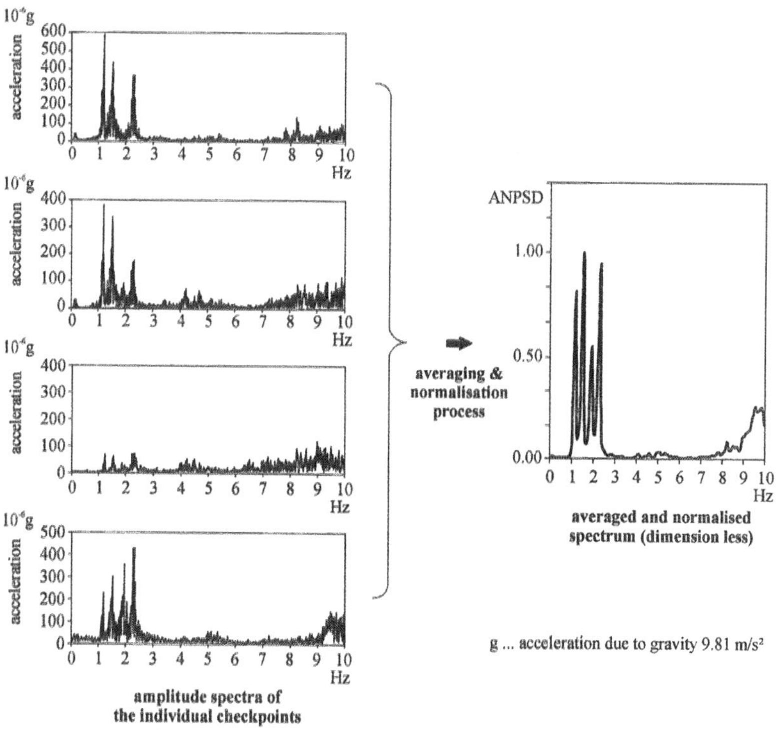

Figure 6.7 Averaged response spectrum of a load-bearing structure

6.5.2 MODE SHAPES

The mode shapes are the vibration forms corresponding to the eigenfrequencies that establish the actual vibrations of the structures. Therefore they are the second essential value for the description of dynamic behaviour of a structure apart from the eigenfrequencies. In every check point the total vibrations of the structure and therefore also the vibration shares of the individual mode shapes are registered.

For the AVM the procedure described in the following was chosen in order to be able to determine quickly the mode shapes from the measured accelerations under ambient excitation. After determination of the eigenfrequencies in the ANPSD spectrum the original records are dually integrated, thus converting the accelerations into vibration distances. The data obtained by this procedure are again transformed into the frequency domain by means of the FFT and the obtained distance spectra are standardized. Now the standardized displacements of the mode shapes can be directly read at the individual check points for every eigenfrequency and be applied via structure geometry by considering the phase information from the FFT (Figure 6.8).

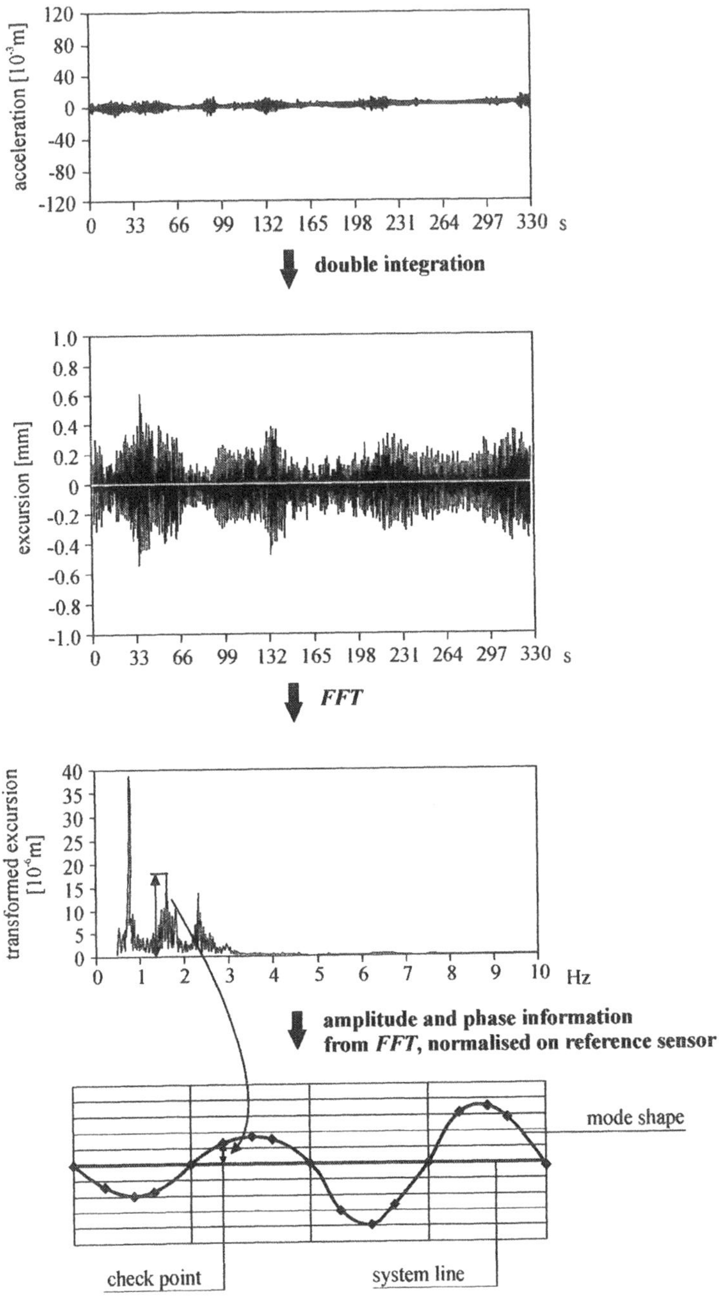

Figure 6.8 Determination of mode shapes

6.5.3 *DAMPING*

Apart from the eigenfrequencies and mode shapes the damping coefficients belonging to the frequencies represent the third value definable from the acceleration measurements. The frequency-dependent damping properties are essential parameters for the assessment of the condition of a structure as the damping coefficients considerably rise with increasing exploitation of the maximum ultimate limit strength, i.e. at the transition from the elastic into the elastoplastic range [5].

For determination of damping from the acceleration records the RDT (Figure 6.9) developed by *NASA* in the 1970s is used. The method is described in detail in reference [6]; at this point only the basic principle is explained.

Averaged time windows from the measured response of a stochastically stressed system only have pure system properties, cleaned from the influences of an accidental stress. Only those time windows fulfilling a special trigger condition are used for averaging. The trigger condition has to be fixed for every individual case and depends on the characteristics of the measured signal. In the course of development of the AVM it turned out that the following averaging of all discrete measuring values x_i, $i = 1, \ldots, n$, is reasonable as the condition

$$x_{\text{Trigger}} = \left(\frac{1}{n} \sum_{i=1}^{n} x_i^2 \right)^{\frac{1}{2}} \tag{6.5}$$

For practical determination of damping relevant to an eigenfrequency the procedure described in Figure 6.9 is chosen for the AVM. Due to uncertainties during the determination the damping coefficient is not only calculated from two neighbouring vibration maximums but different peaks according to the algorithm stated are combined and the damping constants determined from every combination are statistically examined. As a result of these examinations the damping constants corresponding to the eigenfrequencies exist in the form of mean values and the corresponding standard deviations or variations.

6.6 THEORY ON CABLE FORCE DETERMINATION

6.6.1 *FREQUENCIES OF CABLES AS A FUNCTION OF THE INHERENT TENSILE FORCE*

Cables are special one-dimensional structures with little bending stiffness, which carry tangential and lateral loads. Therefore, cables are being stressed in tension only. In the following analysis the linear theory of the free vibration

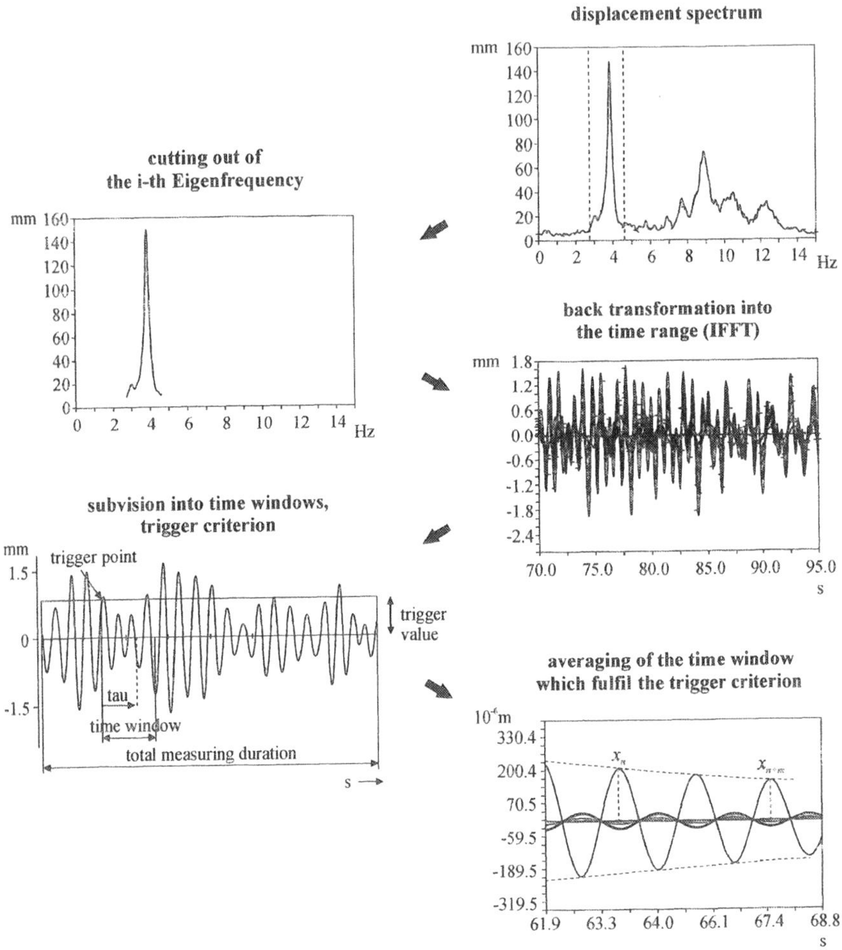

$$\zeta = \frac{1}{m \cdot 2\pi} \cdot \ln\left(\frac{x_n}{x_{n+m}}\right)$$

ζ ... damping coefficient of the i-th eigenfrequency

Figure 6.9 Determination of damping by means of the RDT

of cables is considered with the theoretical assumption that the cable does not have bending stiffness. Additionally, small cable sag and hinged support conditions are assumed.

The equilibrium of a suspended cable with regard to the centre of gravity is considered. The dynamic equilibrium for an infinite small element with length dx is set up according to Figure 6.10. First, the law of conservation of mass is used:

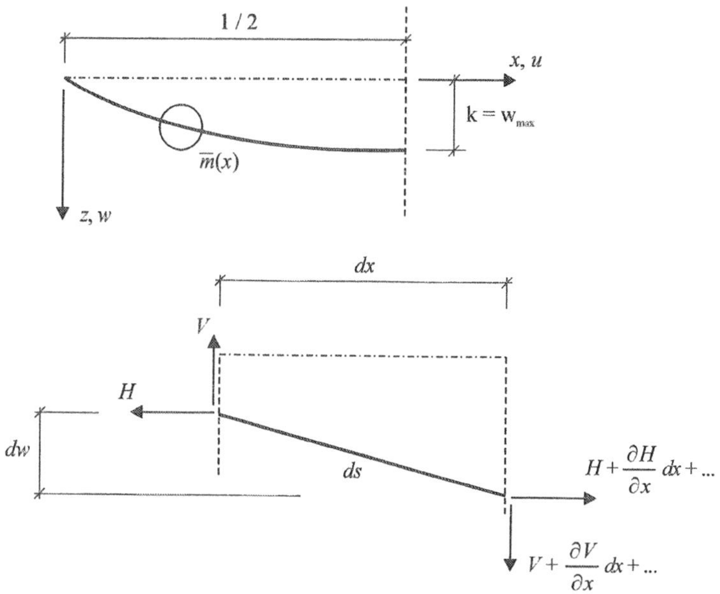

Figure 6.10 The dynamic equilibrium for a suspended cable

$$-V(x,t) + \left[V(x,t) + \frac{\partial V(x,t)}{1!\partial x} dx + \frac{\partial V^2(x,t)}{2!\partial x^2} dx^2 + \frac{\partial V^3(x,t)}{3!\partial x^3} dx^3 + \cdots + \frac{\partial V^n(x,t)}{n!\partial x^n} dx^n + \cdots \right] - f_1(x,t)dx = 0 \tag{6.6}$$

where (in accordance with d'Alembert's principle)

$$f_1(x,t) = \overline{m}(x) \frac{\partial^2 w(x,t)}{\partial t^2} \tag{6.7}$$

is the inertial force. The higher-order Taylor terms vanish immediately after dividing equation (6.6) by dx and are followed by applying the limit d$x \to 0$. This gives

$$\frac{\partial V(x,t)}{\partial x} - \overline{m}(x) \frac{\partial^2 w(x,t)}{\partial t^2} = 0 \tag{6.8}$$

Using the relationship

$$\frac{V(x,t)}{H(x,t)} = \frac{\partial w(x,t)}{\partial x} \tag{6.9}$$

and considering no varying forces in the time domain and remembering that the internal tensile force is constant with respect to the cable length leads to equilibrium of the cable in the in-plane direction after substituting into equation (6.8) as follows:

$$H\frac{\partial^2 w(x,t)}{\partial x^2} = \overline{m}(x)\frac{\partial^2 w(x,t)}{\partial t^2} \tag{6.10}$$

Separating the displacement function by means of the Bernoulli approach,

$$w(x,t) = \sum_{n=1}^{N} \phi_n(x) Z_n(t) \tag{6.11}$$

where $\phi_n(x)$ is the nth mode function and $Z_n(t)$ represents the nth modal coordinates, leads to the following expression for the nth mode:

$$H\phi(x)_{,xx}Z(t) - \overline{m}(x)\phi(x)\ddot{Z}(t) = 0 \tag{6.12}$$

The modified form of this equation

$$-\frac{\ddot{Z}(t)}{Z(t)} = -\frac{H\phi(x)_{,xx}}{\overline{m}(x)\phi(x)} \equiv \omega^2 \tag{6.13}$$

which is substituted to ω^2, since the equation above is valid for any arbitrary t and x, gives the vibrations equation for the modal coordinates as

$$\ddot{Z}(t) + \omega^2 Z(t) = 0 \tag{6.14}$$

The solution for this homogeneous differential equation is given as

$$Z(t) = Z(0)\cos(\omega t) + \frac{\dot{Z}(0)}{\omega}\sin(\omega t) \tag{6.15}$$

where $Z(0)$ and $\dot{Z}(0)$ are the initial conditions. Substituting ω^2 into equation (6.12) gives

$$H\phi(x)_{,xx} + \omega^2\overline{m}(x)\phi(x) = 0 \tag{6.16}$$

with regards to the boundary conditions, the natural modes may be represented in the form

$$\phi_k(x) = C\sin(\lambda_k x) \tag{6.17}$$

with

$$\lambda_k = \frac{k\pi}{l} \tag{6.18}$$

Substitution of equation (6.17) into equation (6.16),

$$-HC\lambda^2 \sin(\lambda x) + \omega^2 \overline{m}(x) C \sin(\lambda x) = 0 \tag{6.19}$$

leads to a second-order polynomial equation with respect to the characteristic roots λ^2:

$$-H\lambda^2 + \omega^2 \overline{m}(x) = 0 \tag{6.20}$$

The final solution with respect to ω_n^2 is obtained by considering the relationship equation (6.18):

$$-H\left(\frac{k\pi}{l}\right)^2 + \omega_k^2 \overline{m}(x) = 0 \Rightarrow \omega_k^2 = \frac{H}{\overline{m}(x)}\left(\frac{k\pi}{l}\right)^2 \tag{6.21}$$

Finally, using the relationship between the natural linear frequency f and the natural circular frequency ω,

$$f = \frac{\omega}{2\pi} \tag{6.22}$$

the equation for the vertical vibration frequency can be expressed as

$$f_k = \frac{k}{2l}\left(\frac{H}{\overline{m}(x)}\right)^{\frac{1}{2}} \tag{6.23}$$

The following relationship exists between the horizontal force component H and the cable force N (Figure 6.10):

$$N\frac{\mathrm{d}x}{\mathrm{d}s} = H \tag{6.24}$$

Furthermore, if the sag of a cable is negligible or small enough (e.g. stay cables and tendons),

$$\mathrm{d}x \approx \mathrm{d}s \tag{6.25}$$

Therefore

$$H \approx N \tag{6.26}$$

Finally, substituting equation (6.26) into equation (6.23) leads to

$$\boxed{f_k = \frac{k}{2l}\left(\frac{N}{\overline{m}(x)}\right)^{\frac{1}{2}}} \tag{6.27}$$

Note, as mentioned before, that equation (6.27) was derived under the assumption of no bending stiffness. This theoretical case will be called the taut string in this section and can be distinguished by the letter 's' as the second letter in the index. Furthermore, the distributed mass is termed m (cable mass per unit for simpler applications):

$$f_{ks} = \frac{k}{2l}\left(\frac{N}{m}\right)^{\frac{1}{2}} \tag{6.28}$$

In a taut string there is a harmonic succession of every mode shape, starting from the first up to very high eigenfrequencies. Figure 6.11 illustrates the harmonic mode shapes of the first five eigenfrequencies.

Since real cables and tendons that are used for civil engineering structures have considerable bending stiffness, cable sag and in most cases a fixed support condition, a method was developed to eliminate the influence of these physical properties from the dynamic parameters obtained using the AVM. This method was developed within the European project IMAC dedicated especially to cable assessment on the basis of the AVM, which was coordinated by the authors.

In section 6.6.5 there is an explanation of the theory behind the method for the adjustments of the dynamic parameters. The impacts on the dynamic parameters owing to the bending stiffness and the support conditions are explained in sections 6.6.2 and 6.6.3 respectively.

6.6.2 *INFLUENCE OF THE BENDING STIFFNESS*

A real cable deviates from the linear relation known from the theoretical case of taut string since the bending stiffness causes an increase of eigenfrequencies in the higher modes due to the bending stiffness. Figure 6.12 shows the course of the relation between the eigenfrequencies (Hz) and the respective order of the eigenfrequency. The eigenfrequencies of taut string aligns with the linear line whereas the eigenfrequencies of a real cable are noticeable higher than values from a linear relation. Due to the low modal curvature in the fundamental mode shapes (f_1 to f_5) during the vibrations, bending stiffness has only minor effects on the first five eigenfrequencies but the influence increases continuously in the higher ones. This coherence of stiffness and deviation of eigenfrequency

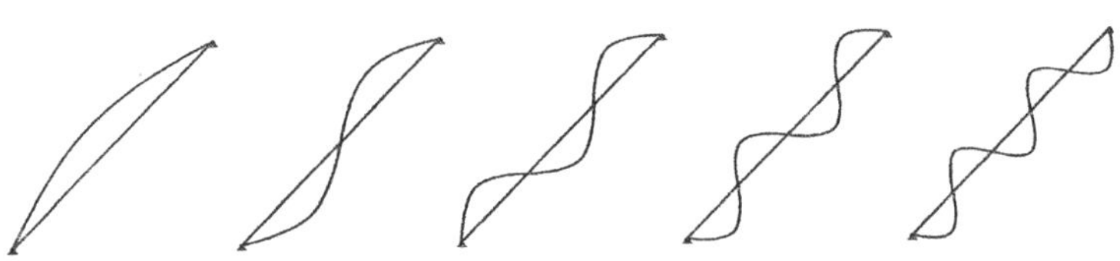

Figure 6.11 Mode shapes f_1 to f_5

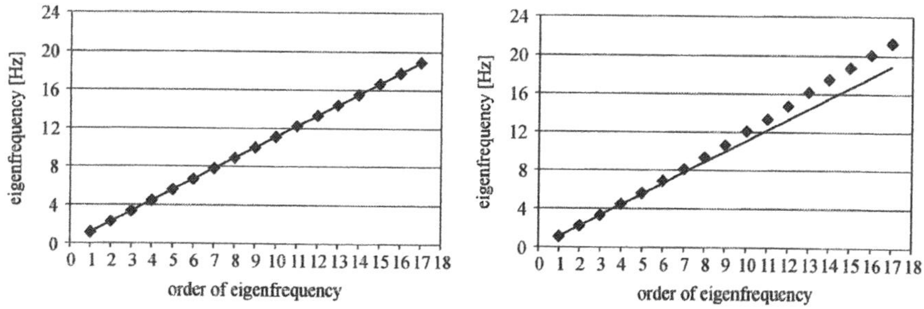

Figure 6.12 Frequency progress of a taut string (left) and a real cable (right)

from a linear behaviour expresses the necessity to consider the bending stiffness for accurate cable force calculation [7].

For the theoretical investigations of the bending stiffness two different cases are defined using the cable theory for the theoretical case of taut string and the beam theory, which considers the bending stiffness for a real cable. Both cases have hinged support conditions defined (Figure 6.13). Borderline case 1 allows

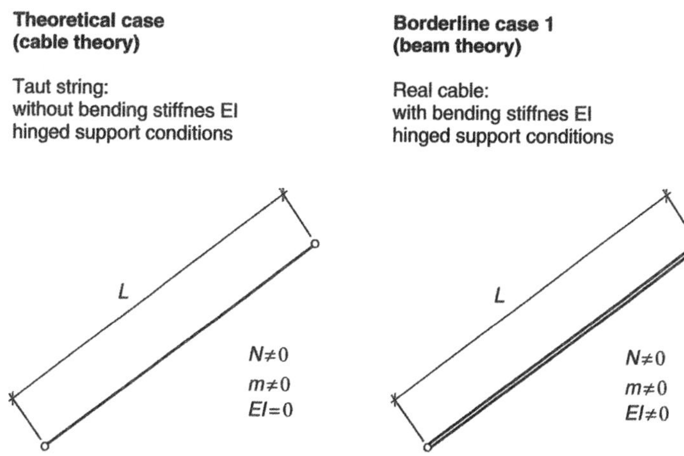

$$f_{ks} = \frac{k}{2l}\left(\frac{N}{m}\right)^{\frac{1}{2}} \quad (6.28)$$

$$f_k = \frac{k}{2l}\left(\frac{N}{m}\right)^{\frac{1}{2}}\left[1 + \left(\frac{k\pi}{\xi}\right)^2\right]^{\frac{1}{2}} \quad (6.29)$$

with

$$\xi = l\left(\frac{N}{EI}\right)^{\frac{1}{2}} \quad (6.30)$$

Figure 6.13 Influence of bending stiffness in modal analyses of cables [8]

accurate calculation of the dynamic behaviour of a real cable with hinged support conditions by the parameters length (m), tension N (kN), specific mass m (kg/m) and bending stiffness EI (N m^2).

6.6.3 *INFLUENCE OF THE SUPPORT CONDITIONS*

Unlike the cable theory a definition of the support conditions is required for the beam theory since fixed support conditions become effective for calculations that consider the bending stiffness. In the case of a real cable with fixed support conditions, the free vibration length of a cable does not correspond to the distance between the bearings, as in the case of hinged support conditions, but is shorter. The free vibration length for fixed support conditions is termed the effective length.

Figure 6.14 shows the course of a cable with the same geometry and material properties but different boundary conditions: hinged support conditions for the left graph and fixed support conditions for the right graph. Both graphs show the same curvature in the course of the eigenfrequencies, which is due to the bending stiffness. However, the right graph also deviates from the linear relation known from the taut string, which means a shift upwards.

For further studies on the influence of the boundary condition borderline case 1 was extended with borderline case 2, where fixed support conditions were defined and the beam theory was also applied. The cases are termed 'borderline' cases since, with respect to the stiffness of the anchorage system, the definition 'fixed support conditions' is appropriate for external tendons. For stay cables, where the anchorage system cannot be regarded as totally stiff and motionless, a partial fixed support condition might be correct. This means that the boundary conditions lie between borderline cases 1 and 2. Borderline case 2 allows accurate calculation of the dynamic behaviour for a real cable with fixed support conditions using the parameters length (m), tension N (kN), specific mass m (kg/m) and bending stiffness EI (N m^2) (Figure 6.15).

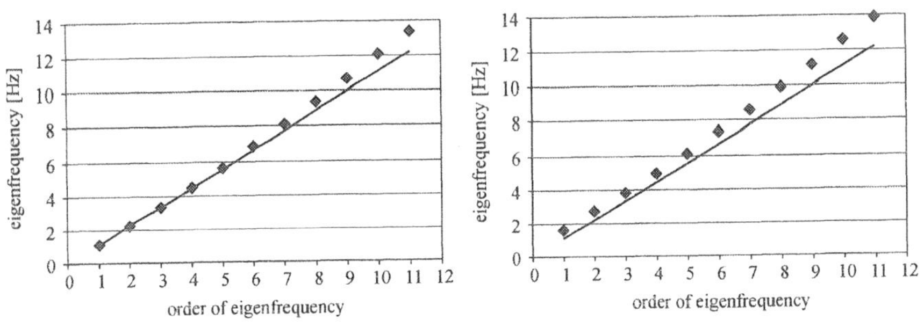

Figure 6.14 Frequency progress of a hinged (left) and a fixed end (right) cable considering the stiffness

Borderline case 1 (beam theory)

Real cable:
with bending stiffness EI
hinged support conditions

Borderline case 2 (beam theory)

Real cable:
with bending stiffness EI
fixed support conditions

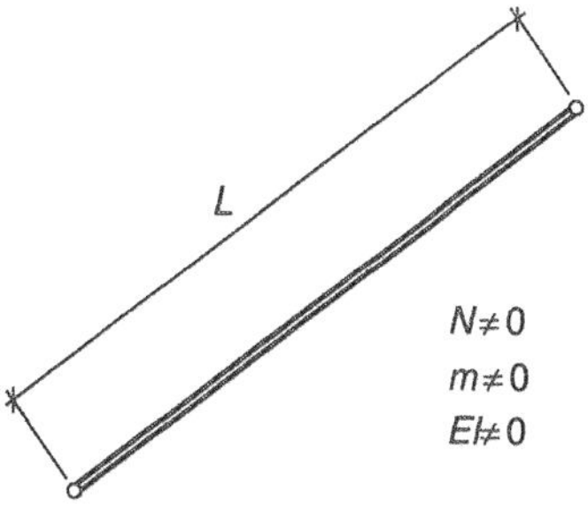

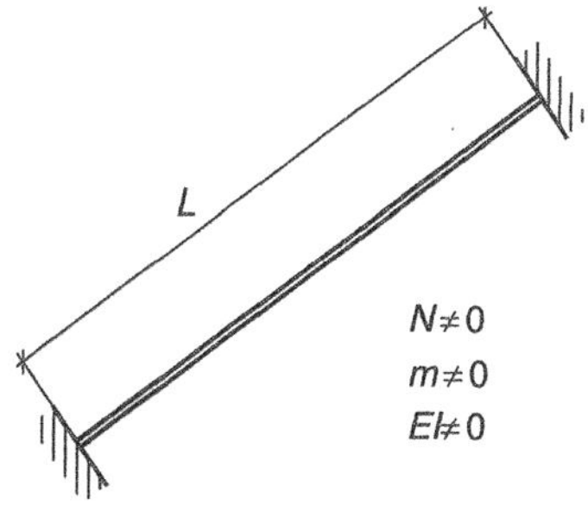

$$f_k = \frac{k}{2l}\left(\frac{N}{m}\right)^{\frac{1}{2}}\left[1+\left(\frac{k\pi}{\xi}\right)^2\right]^{\frac{1}{2}} \quad (6.29)$$

with

$$\xi = l\left(\frac{N}{EI}\right)^{\frac{1}{2}} \quad (6.30)$$

$$f_k = \frac{k}{2l}\left(\frac{N}{m}\right)^{\frac{1}{2}}\left[1+\frac{2}{\xi}+\left(4+\frac{k^2\pi^2}{2}\right)\frac{1}{\xi^2}\right] \quad (6.31)$$

with

$$\xi = l\left(\frac{N}{EI}\right)^{\frac{1}{2}} \quad (6.30)$$

Figure 6.15 Borderline cases of support conditions for modal analyses [8]

6.6.4 COMPARISON OF THE DEFINED CASES WITH EXPERIMENTAL RESULTS

The equations from the cases defined in sections 6.6.2 and 6.6.3 were used to calculate a course of eigenfrequencies in relation to the respective order of eigenfrequencies for verification of analytical studies. Therefore cable characteristics from a tendon (Table 6.1) were used from which dynamic parameters, well-known boundary conditions and the actual cable force measured by load cells is available.

Table 6.1 Cable parameters

Cable length	l	18.54	m
Specific cable mass	m	6.125	kg/m
Cable force	N	100.08	kN
Bending stiffness	EI	1.77	kN m^2

In addition to the equations the same cable characteristics were used to generate a course in an FEM simulation software (ANSYS) as well as in a three-dimensional framework analysis package (RSTAB). The results of the three kinds of generated data as well as eigenfrequencies acquired using the AVM are listed in Table 6.2.

The following is a definition of case numbers:

- Theoretical case: cable theory, without bending stiffness, hinged supported
- Borderline case 1: beam theory, with bending stiffness, hinged supported
- Borderline case 2: beam theory, with bending stiffness, fixed supported

For better differentiation the differences of succeeding eigenfrequencies but not the absolute eigenfrequencies are illustrated in Figures 6.16 to 6.18, which are listed in Table 6.3.

Figures 6.16 to 6.18 illustrate the influence of the bending stiffness, which is clearly visible. The theoretical case based on the cable theory shows a constant difference of succeeding eigenfrequencies, which is due to disregard of the bending stiffness. Borderline cases 1 and 2, which are based on the beam theory and consider the stiffness of the cable, show a progressive course in the sequel

Table 6.2 Generated eigenfrequencies

Case number	Kind of calculation									
	AVM	Analytical			RSTAB			ANSYS		
		0	1	2	0	1	2	0	1	2
f_1	3.56	3.45	3.448	3.50	3.45	3.45	3.50	3.45	3.45	3.50
f_2	7.09	6.89	6.90	7.00	6.89	6.90	7.00	6.89	6.90	7.00
f_3	10.64	10.34	10.37	10.52	10.34	10.37	10.52	10.34	10.37	10.52
f_4	14.21	13.79	13.85	14.05	13.79	13.85	14.05	13.79	13.85	14.05
f_5	17.80	17.24	17.35	17.6	17.24	17.35	17.60	17.24	17.35	17.60
f_6	21.42	20.68	20.87	21.17	20.68	20.87	21.18	20.68	20.87	21.18
f_7	25.04	24.13	24.43	24.78	24.13	24.43	24.79	24.13	24.43	24.79
f_8	28.74	27.58	28.02	28.43	27.58	28.02	28.44	27.58	28.02	28.43
f_9	32.39	31.03	31.66	32.12	31.03	31.66	32.12	31.03	31.66	32.12
f_{10}	36.19	34.47	35.34	35.85	34.47	35.34	35.86	34.48	35.34	35.85
f_{11}	39.99	37.92	39.07	39.64	37.92	39.07	39.64	37.92	39.07	39.64
f_{12}	43.95	41.37	42.85	43.48	41.37	42.85	43.49	41.37	42.85	43.48
f_{13}	47.79	44.82	46.7	47.39	44.82	46.70	47.39	44.82	46.70	47.38
f_{14}	51.77	48.26	50.61	51.37	48.26	50.61	51.36	48.27	50.61	51.35
f_{15}	55.77	51.71	54.58	55.42	51.71	54.59	55.39	51.72	54.58	55.38
f_{16}	59.93	55.16	58.63	59.54	55.16	58.63	59.50	55.17	58.63	59.49
f_{17}	64.02	58.60	62.76	63.76	58.61	62.76	63.76	58.62	62.75	63.68
f_{18}	68.37	62.05	66.96	68.06	62.05	66.97	67.95	62.06	66.96	67.94
f_{19}	72.60	65.50	71.25	72.46	65.50	71.25	72.31	65.51	71.24	72.29
f_{20}	77.06	68.95	75.63	76.95	68.95	75.63	76.75	68.96	75.62	76.73

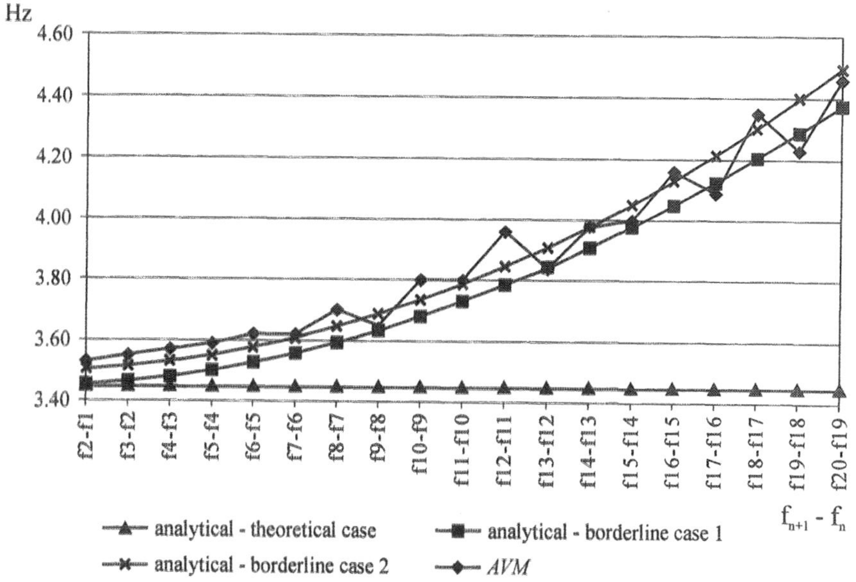

Figure 6.16 Differences of succeeding eigenfrequencies (data from Table 6.3: analytical)

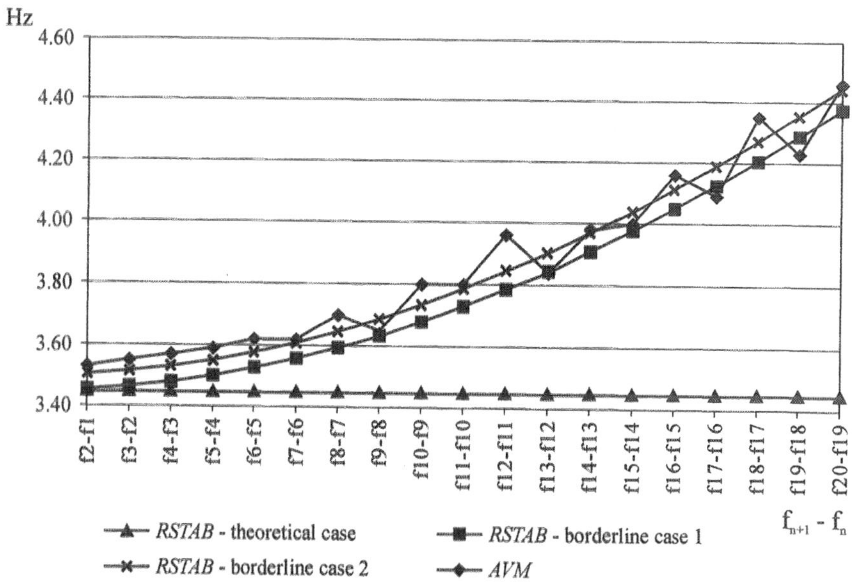

Figure 6.17 Differences of succeeding eigenfrequencies (data from Table 6.3: RSTAB)

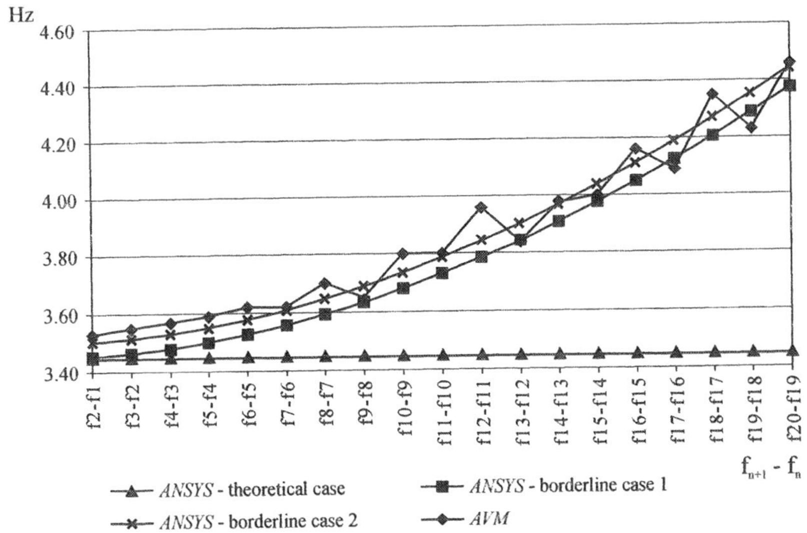

Figure 6.18 Differences of succeeding eigenfrequencies (data from Table 6.3: ANSYS)

Table 6.3 Differences of succeeding eigenfrequencies

Case number	Kind of calculation									
	i54	Analytical			RSTAB			ANSYS		
		0	1	2	0	1	2	0	1	2
f_2–f_1	3.53	3.45	3.45	3.50	3.45	3.45	3.50	3.45	3.45	3.50
f_3–f_2	3.55	3.45	3.46	3.51	3.45	3.46	3.51	3.45	3.46	3.51
f_4–f_3	3.57	3.45	3.48	3.53	3.45	3.48	3.53	3.45	3.48	3.53
f_5–f_4	3.59	3.45	3.50	3.55	3.45	3.50	3.55	3.45	3.50	3.55
f_6–f_5	3.62	3.45	3.53	3.58	3.45	3.53	3.58	3.45	3.53	3.58
f_7–f_6	3.62	3.45	3.56	3.61	3.45	3.56	3.61	3.45	3.56	3.61
f_8–f_7	3.70	3.45	3.59	3.65	3.45	3.59	3.65	3.45	3.59	3.65
f_9–f_8	3.65	3.45	3.63	3.69	3.45	3.63	3.69	3.45	3.63	3.69
f_{10}–f_9	3.80	3.45	3.68	3.73	3.45	3.68	3.73	3.45	3.68	3.73
f_{11}–f_{10}	3.80	3.45	3.73	3.79	3.45	3.73	3.79	3.45	3.73	3.79
f_{12}–f_{11}	3.96	3.45	3.79	3.84	3.45	3.79	3.84	3.45	3.79	3.84
f_{13}–f_{12}	3.84	3.45	3.84	3.91	3.45	3.85	3.90	3.45	3.84	3.90
f_{14}–f_{13}	3.98	3.45	3.91	3.98	3.45	3.91	3.97	3.45	3.91	3.97
f_{15}–f_{14}	4.00	3.45	3.98	4.05	3.45	3.98	4.04	3.45	3.98	4.03
f_{16}–f_{15}	4.16	3.45	4.05	4.13	3.45	4.05	4.11	3.45	4.05	4.11
f_{17}–f_{16}	4.09	3.45	4.12	4.21	3.45	4.13	4.19	3.45	4.12	4.18
f_{18}–f_{17}	4.35	3.45	4.20	4.30	3.45	4.21	4.27	3.45	4.20	4.27
f_{19}–f_{18}	4.23	3.45	4.29	4.40	3.45	4.29	4.35	3.45	4.29	4.35
f_{20}–f_{19}	4.46	3.45	4.38	4.50	3.45	4.38	4.44	3.45	4.37	4.44

of the eigenfrequency differences. The difference between borderline cases 1 and 2 (the shift of eigenfrequencies moves upwards), which was explained in the preceding section, can be found in all diagrams.

Besides the course of the calculated and generated eigenfrequencies, the same dynamic parameter for an external tendon received by the AVM is present in every diagram where this eigenfrequency fits borderline case 2 very well.

The high affinity of the data in each single case independent of its source (analytical, RSTAB and ANSYS), verifies the validity of the equations defined in sections 6.6.2 and 6.6.3.

6.6.5 *MEASUREMENT DATA ADJUSTMENT FOR EXACT CABLE FORCE DETERMINATION*

For determination of the accurate cable the equation of borderline case 2 is predestined for practical application because it fits the dynamic behaviour of a real cable [8].

Borderline case 2:

$$f_k = \frac{k}{2l}\left(\frac{N}{m}\right)^{\frac{1}{2}}\left[1 + \frac{2}{\xi} + \left(4 + \frac{k^2\pi^2}{2}\right)\frac{1}{\xi^2}\right] \tag{6.31}$$

Equation parameters:
Known from AVM measurement:
f_k = eigenfrequency of the kth order (Hz)
k = order of eigenfrequency

Known from design documents:
l = cable length (m)
m = cable mass per unit length (kg/m)

Unknown parameters in the equation of borderline case 2:
ξ = related bending stiffness
N = cable force (kN)

Since the related bending stiffness ξ is not known for cables or tendons, the cable force has to be determined by the AVM, which was developed to eliminate the influence of the bending stiffness within the European project IMAC. Elimination of the influence of bending stiffness means that the eigenfrequency from the measurement f_k is reduced to the theoretical eigenfrequency f_{ks}, which corresponds to a cable with the same geometry without any bending stiffness [8,9].

Therefore the authors developed a software program with the aim to determine the exact cable force without having information about the bending stiffness. The idea behind this software is the adjustment of the measured nonlinear relation between eigenfrequencies and their order to a linear relation such as in the theoretical case (taut string).

From an analytical point of view the parameters $f_{ks}(k)$ and β_k (k, ξ) in equation (6.32) have to be varied until the sum of squares of the difference measured f_k and calculated f_k is a minimum. The fitting procedure uses non-linear least squares data of the Gauss–Newton method [9,10]:

$$f_k = \underbrace{\frac{k}{2l}\left(\frac{N}{m}\right)^{\frac{1}{2}}}_{f_{ks}} \underbrace{\left[1 + \frac{2}{\xi} + \left(4 + \frac{k^2\pi^2}{2}\right)\frac{1}{\xi^2}\right]}_{\beta_k} \Rightarrow f_k = f_{ks}(k)\ \beta_k(k, \xi) \qquad (6.32)$$

Figure 6.19 represents the surface of the sum of squares of the difference measured and calculated eigenfrequencies. At the point of minimum the parameters f_{1s} and ξ have their best adjustment and f_{1s} corresponds to the first eigenfrequency of the cable if there were not any bending stiffness. The eigenfrequency f_{1s} can be used to calculate the accuracy of the cable force with the equation of the theoretical case 'taut string':

$$f_{ks} = \frac{k}{2l}\left(\frac{N}{m}\right)^{\frac{1}{2}} \qquad (6.28)$$

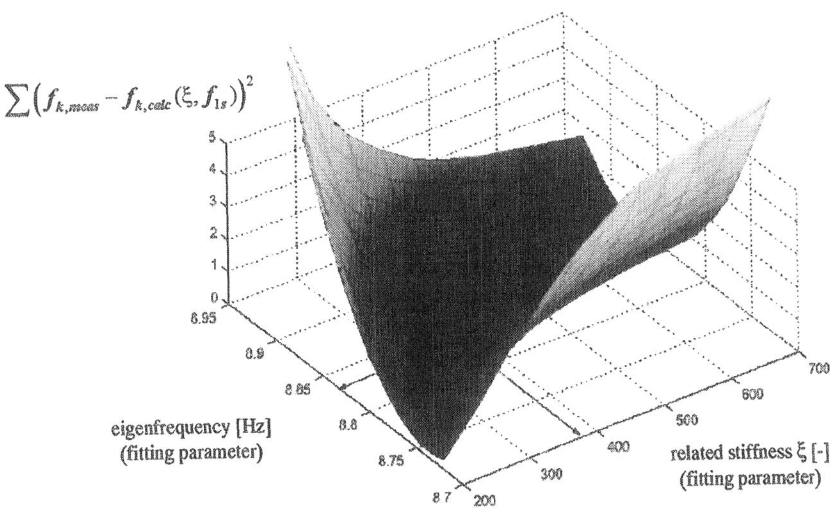

Figure 6.19 Search for the minimum in the deviation

6.7 TRANSFER FUNCTIONS ANALYSIS

This subsection has three main purposes. The first purpose is to collect in one document the various transfer functions with regard to the mathematics. The second purpose is to present the state-of-the-art in terms of practical applications, e.g. mechanical, electrical engineering, respectively, etc. The third purpose is to give some outlook on future works.

6.7.1 MATHEMATICAL BACKGROUNDS [11]

From a mathematical point of view, any transfer function is a mathematical statement that describes the transfer characteristics of a system, subsystem or equipment. In general, each transfer function can be represented as the relationship between the input and the output of a system:

$$\mathrm{Out[put]} = H(f) \times \mathrm{In[put]} \tag{6.33}$$

In mathematics, there are three well-known cases of transformation where the transform function is needed: Fourier, Laplace and Z-Transformations.

6.7.1.1 Fourier Transformation

Periodic functions may be represented with very good accuracy by means of the Fourier transformation. The main characteristic of a periodic function is the fact that it can be displaced by any whole-number multiple of its period without values changing, e.g.

$$f(t + nT) = f(t), \qquad n = 0, \pm 1, \pm 2, \pm 3, \ldots \tag{6.34}$$

Figure 6.20 depicts some typical periodic functions. However, two forms of this transformation are available, Fourier series and Fourier integrals. The Fourier series technique uses the expansion of 2π-periodic, piecewise monotone and continuous functions into series of trigonometric functions:

$$f(t) = \frac{a_0}{2} + \sum_{n=1}^{\infty} [a_n \cos(nt) + b_n \sin(nt)] \tag{6.35}$$

where the Fourier coefficients a_n and b_n are given as follows:

$$\begin{aligned} a_n &= \frac{1}{\pi} \int_{-\pi}^{\pi} f(t) \cos(nt) \mathrm{d}t, \qquad n = 0, 1, 2, \ldots \\ b_n &= \frac{1}{\pi} \int_{-\pi}^{\pi} f(t) \sin(nt) \mathrm{d}t, \qquad n = 1, 2, 3, \ldots \end{aligned} \tag{6.36}$$

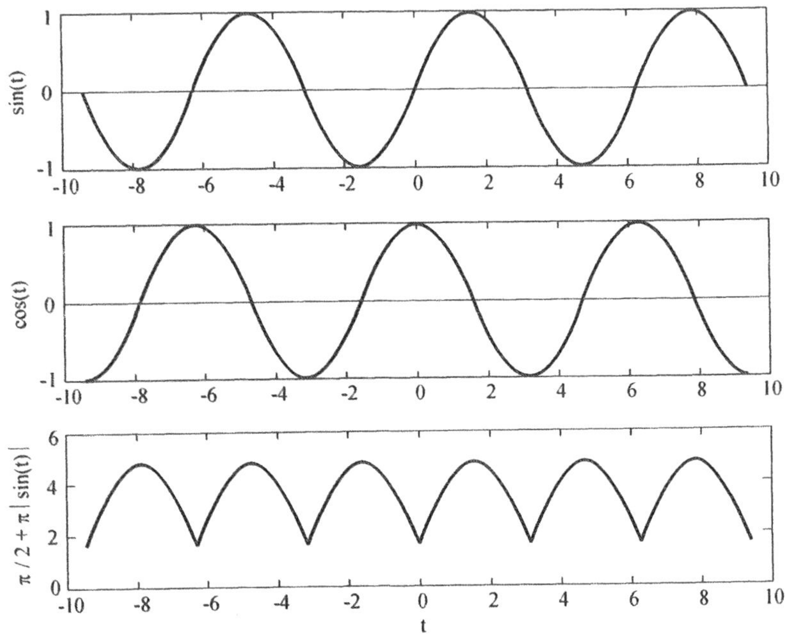

Figure 6.20 Typical period functions

For instance, the sawtooth function (Figure 6.21) is given by the following expression:

$$f(t) = 2\left[\frac{\sin(t)}{1} - \frac{\sin(2t)}{2} + \frac{\sin(3t)}{3} - \cdots\right] \tag{6.37}$$

where $a_0 = a_n = 0$ and $b_n = (-1)^{n+1}\, 2/n$ for $n = 1, 2, 3, \ldots$. However, the limitation $[-\pi,\pi]$ can be abolished by introduction of the so-called angular frequency $\omega = 2\pi/T$ and by the substitution $t \to \omega t$. Therefore, the Fourier series expression (equation (6.34)) can be rewritten as

$$f(t) = \frac{a_0}{2} + \sum_{n=1}^{\infty}[a_n \cos(n\omega t) + b_n \sin(n\omega t)] \tag{6.38}$$

where

$$\begin{aligned} a_n &= \frac{2}{T}\int_{-T/2}^{T/2} f(t)\cos(n\omega t)\mathrm{d}t, \qquad n = 0, 1, 2, \ldots \\ b_n &= \frac{2}{T}\int_{-T/2}^{T/2} f(t)\sin(n\omega t)\mathrm{d}t, \qquad n = 1, 2, 3, \ldots \end{aligned} \tag{6.39}$$

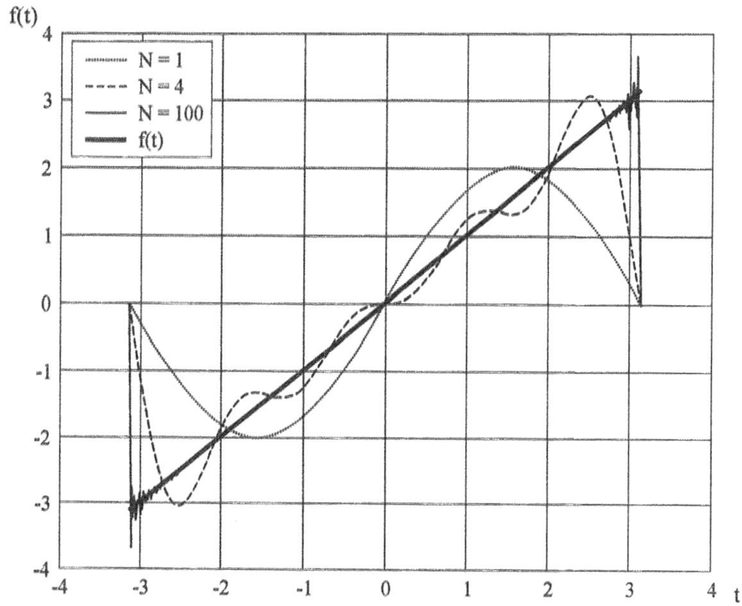

Figure 6.21 Sawtooth function within the interval $[-\pi,\pi]$ and its extension by means of the Fourier series with $N=1$, $N=4$ and $N=100$ ($N=n_{\max}$)

Figure 6.22 gives a view of the stepping function

$$f(t) = \begin{cases} C, & 0 < t \le T/2 \\ -C, & -T/2 \le t < 0 \end{cases} \tag{6.40}$$

Its Fourier series expression for a period T is given by

$$f(t) = \frac{4C}{\pi}\left(\frac{\sin(1\,\omega t)}{1} + \frac{\sin(3\,\omega t)}{3} + \frac{\sin(5\,\omega t)}{5} + \cdots\right) \tag{6.41}$$

where the coefficients $a_0 = a_n = b_{2n} = 0$ and

$$b_{2n-1} = \frac{4C}{\pi(2n-1)}, \qquad n = 1,2,\ldots \tag{6.42}$$

Note that Dirichlet's conditions must be fulfilled in order to apply the Fourier series.

As a generalization of the Fourier series the Fourier integrals have an essential characteristic, namely the ability to represent not only periodic functions but non-periodic ones as well. For instance, differential equations often become straightforward to be solved by means of algebraic transformations.

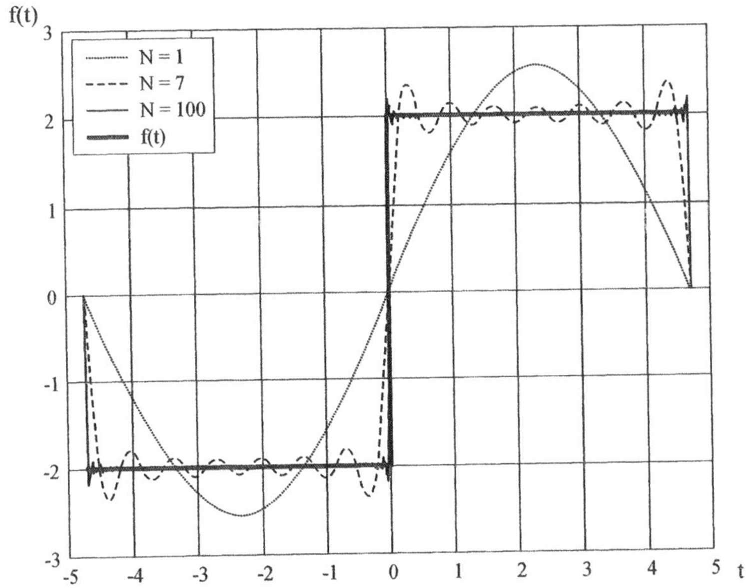

Figure 6.22 Stepping function within the interval $[-T/2, T/2]$ with $T = 3\pi$ and $C = 2$, and its extension by means of the Fourier series with $N = 1$, $N = 7$ and $N = 100$ ($N = n_{\max}$)

The Fourier integral, an integral expansion of non-periodic functions by means of trigonometric functions (continuous spectra), can be obtained by shifting the interval limits to infinity ($T \to \pm\infty$):

$$f(t) = \int_0^{\infty} [a(\omega)\cos(\omega t) + b(\omega)\sin(\omega t)]\mathrm{d}\omega \tag{6.43}$$

where the coefficients can be written by

$$\begin{aligned} a(\omega) &= \frac{1}{\pi}\int_{-\infty}^{\infty} f(\tau)\cos(\omega\tau)\mathrm{d}\tau \\ b(\omega) &= \frac{1}{\pi}\int_{-\infty}^{\infty} f(\tau)\sin(\omega\tau)\mathrm{d}\tau \end{aligned} \tag{6.44}$$

Dirichlet's conditions shall be fulfilled again in this case. Figure 6.23 shows the change in the periodic stepping function to non-periodic by means of the limit $T \to \infty$.

Next, the complex representation of the Fourier integrals is given by

$$f(t) = \frac{1}{2\pi}\int_{-\infty}^{\infty} F(\omega)\mathrm{e}^{\mathrm{j}\omega t}\mathrm{d}\omega \tag{6.45}$$

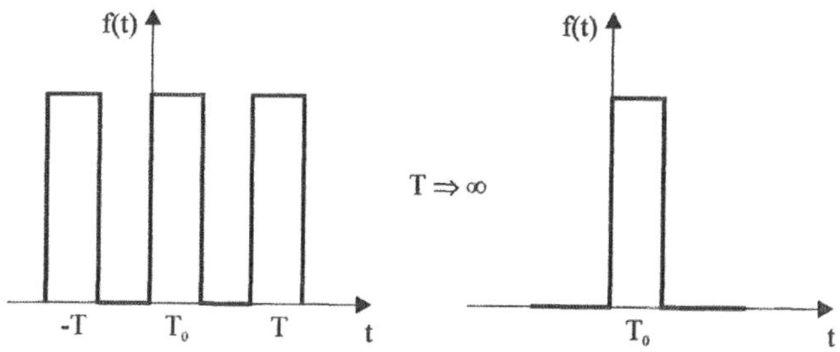

Figure 6.23 Transfer of the stepping function from periodic to non-periodic by means of the limit $T \to \infty$

and

$$F(\omega) = \int_{-\infty}^{\infty} f(t)\mathrm{e}^{-\mathrm{j}\omega t}\mathrm{d}t \tag{6.46}$$

where $F(\omega) \hat{=} F(f(t))$ is the so-called Fourier transformed function (spectral function) from $f(t)$ and $f(t) = F^{-1}$ $[F(\omega)]$ is the Fourier transformation (operator F). Generally, the Fourier transformation represents a time domain into its corresponding spectral domain by assignment of the function $F(\omega) = F(f(t))$ to the function $f(t)$. The transition from $F(\omega)$ to $f(t)$ is referred to as the inverse Fourier transformation (Operator F^{-1}) because of the symmetry of $f(t)$ and $F(\omega)$. For instance, any even function $f(t) = f(-t)$ is given by the relationship (Figure 6.24)

$$\begin{aligned} f(t) &= \frac{2}{\pi}\int_0^{\infty} F_c(\omega)\cos(\omega t)\mathrm{d}\omega \\ F_c(\omega) &= \int_0^{\infty} f(t)\cos(\omega t)\mathrm{d}t \end{aligned} \tag{6.47}$$

which is also called the Fourier cosine transformation. On the other hand, for odd functions $f(t) = -f(-t)$ the Fourier sine transformation can be used:

$$\begin{aligned} f(t) &= \frac{2}{\pi}\int_0^{\infty} F_s(\omega)\sin(\omega t)\mathrm{d}\omega \\ F_s(\omega) &= \int_0^{\infty} f(t)\sin(\omega t)\mathrm{d}t \end{aligned} \tag{6.48}$$

However, for the purpose of practical applications the discrete Fourier transformation (DFT) is recommended, which is a numerical generalization of equation (6.48) at certain discrete points. Moreover, the fast Fourier transformation (FFT) is applied as a very efficient and accurate evaluation of the DFT.

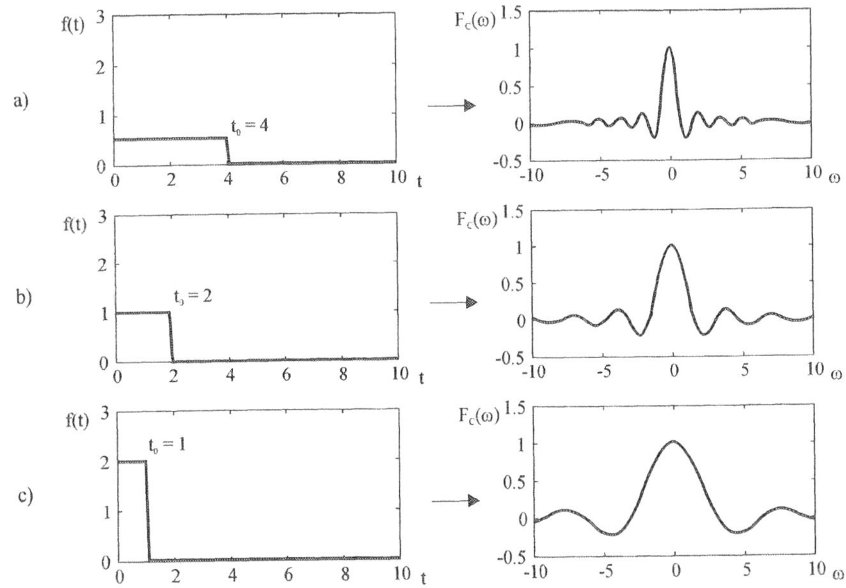

Figure 6.24 Fourier cosine transformation for the stepping function for: (a) $t_0 = 4$; (b) $t_0 = 2$; (c) $t_0 = 1$

6.7.1.2 Laplace and Z-Transformation

The Laplace transformation, $\mathcal{L}\{f(t)\}$, is basically used in order to simplify the solution procedure of linear differential equations with constant coefficients, where instead of the direct solution of a differential equation the solution of an algebraic equation through the image space is taking place. The main phases of the solving procedure can be presented as follows:

- Laplace transformation: transformation of the differential equation into an algebraic equation.
- Algebraic solution through the image space: solution of the algebraic equation with respect to the so-called image function of the solution.
- Inverse transformation: inverse Laplace transformation $\mathcal{L}^{-1}$ of the image function in order to obtain the original function and the final solution of the differential equation through the original domain.

Therefore, the Laplace transformation can also be defined as the assignment of an image function to a time function (original function):

$$\mathcal{L}\{f(t)\} = F(s) \int_0^\infty f(t)e^{-st}\mathrm{d}t \tag{6.49}$$

where the new variable $s=\delta+j\omega$ is complex in general, $s\in C$, and δ and ω are real numbers that define the locations of s in the complex plane.

The following example will now be considered. Determine the water level $h(t)$ in a vessel with a base area of A. The inflow is given by the relation $q(t)=we^{-t}$, whereas the outflow is $kh(t)$. The coefficients are specified as $w=4\,\mathrm{m}^3/\mathrm{s}$ and $k=\frac{1}{2}\,\mathrm{m}^2/\mathrm{s}$, and the initial conditions are given by $h(t=0)=h_0=4\,\mathrm{m}$. In accordance with the law of conservation of mass, the following first-order linear differential equation is obtained:

$$A\frac{\mathrm{d}h(t)}{\mathrm{d}t}+kh(t)=q(t)=we^{-t} \tag{6.50}$$

Transformation of the differential equation into the image domain leads to

$$A[sH(s)-h_0]+kH(s)=Q(s)=\frac{w}{s+1}\Rightarrow\frac{1}{4}[sH(s)-h_0] \\ +\frac{1}{2}H(s)=\frac{4}{s+1} \tag{6.51}$$

Next, the algebraic solution through the image domain gives

$$\frac{1}{4}(s+2)H(s)=\frac{4}{s+1}+1=\frac{s+5}{s+1}$$

$$\Rightarrow \boxed{H(s)=\frac{4(s+5)}{(s+1)(s+2)}=\frac{16}{s+1}-\frac{12}{s+2}} \tag{6.52}$$

Finally, the original solution is obtained by means of inverse transformation into the original domain, namely (Figure 6.25)

$$h(t)=\mathscr{L}^{-1}\left\{\frac{16}{s+1}\right\}-\mathscr{L}^{-1}\left\{\frac{12}{s+2}\right\}=\boxed{16e^{-t}-12e^{-2t}} \tag{6.53}$$

Furthermore, the Z-transformation represents the discrete analogue of the Laplace transformation, similar to the DFT and FFT procedures mentioned in the previous subsection.

6.7.2 *TRANSFER FUNCTIONS IN THE VIBRATION ANALYSIS [12,13]*

The vibration analysis usually deals with the dynamical structural behaviour due to an arbitrary general loading $p(t)$, as illustrated in Figure 6.26. Next,

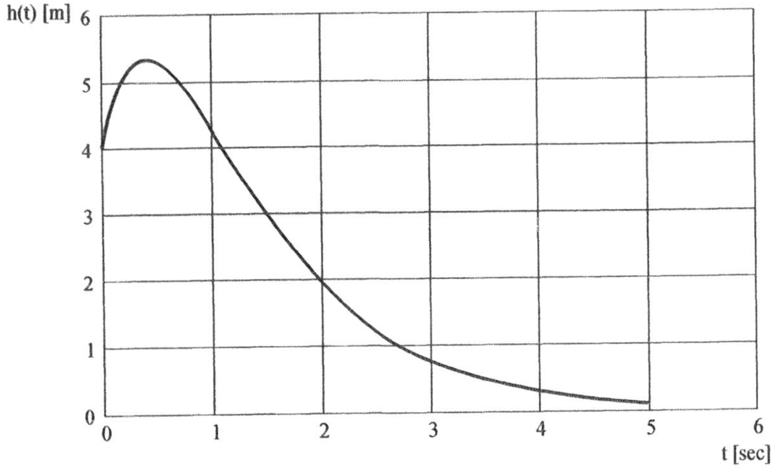

Figure 6.25 Solution for the water level through the time domain

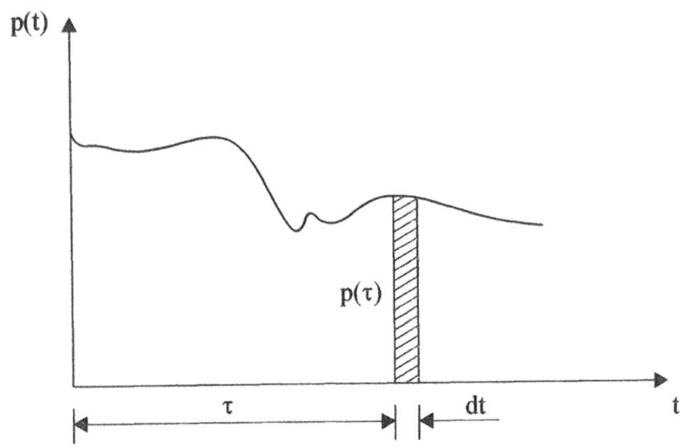

Figure 6.26 Arbitrary general loading

concentrate on the intensity of loading $p(\tau)$ acting at time $t=\tau$. This loading acting during the time interval of time dt represents a very short duration impulse $p(\tau)d\tau$ on the structure. Thus, for the differential time interval $d\tau$, the response produced by the impulse $p(\tau)d\tau$ is exactly equal to

$$dw(t) = \frac{p(\tau)d\tau}{m\omega}\sin\omega(t-\tau), \qquad t \geq \tau \tag{6.54}$$

In this expression, the term $dw(t)$ represents the time history response to the differential impulse over the entire time $t \geq \tau$. The entire history can be considered to consist of a succession of short impulses, each producing its own differential response in the form of equation (6.54). Thus, the total response can be obtained by summing all the differential responses developed during the loading history by integrating equation (6.54) as follows:

$$w(t) = \frac{1}{m\omega}\int_0^t p(\tau)\sin\omega(t-\tau)\mathrm{d}\tau, \qquad t \geq 0 \tag{6.55}$$

This is the so-called Duhamel integral and can be used to evaluate the response of an undamped Single-degree-of-freedom (SDOF) system to any form of dynamic loading $p(t)$. Equation (6.55) can also be expressed in the general convolution integral form:

$$w(t) = \int_0^t p(\tau)h(t-\tau)\mathrm{d}\tau, \qquad t \geq 0 \tag{6.56}$$

in which the function

$$h(t-\tau) = \frac{1}{m\omega}\sin\omega(t-\tau) \tag{6.57}$$

is known as the unit-impulse response function because it expresses the response of the SDOF system to a pure impulse of unit magnitude applied at time $t=\tau$. Note that this approach may be applied only to linear systems because the response is obtained by superposition of individual impulse responses.

Next, the assumption that the loading is initiated at time $t=0$ and that the structure is at rest at that time will be neglected. Thus, for any other specified initial conditions $w(0) \neq 0$ and $\dot{w}(0) \neq 0$, the additional free-vibration response must be added to the solution:

$$w(t) = \frac{\dot{w}(0)}{\omega}\sin\omega t + w(0)\cos\omega t + \frac{1}{m\omega}\int_0^t p(\tau)\sin\omega(t-\tau)\mathrm{d}\tau \tag{6.58}$$

Similarly to equation (6.58), the dynamic response of an under-critically damped SDOF system results in

$$w(t) = \frac{\dot{w}(0)}{\omega_{\mathrm{D}}}\sin\omega_{\mathrm{D}}t + w(0)\cos\omega_{\mathrm{D}}t + \int_0^t p(\tau)h(t-\tau)\mathrm{d}\tau \tag{6.59}$$

where

$$h(t-\tau) = \frac{1}{m\omega_{\mathrm{D}}}\sin\omega_{\mathrm{D}}(t-\tau)\exp[-\zeta\omega(t-\tau)] \tag{6.60}$$

and

$$\omega_{\mathrm{D}} = \omega(1-\zeta^2)^{\frac{1}{2}} \tag{6.61}$$

It is sometimes more convenient, however, to perform the analysis in the frequency domain. Using the Fourier integrals, equations (6.45) and (6.46), the loading can be represented as

$$\begin{aligned} p(t) &= \frac{1}{2\pi}\int_{-\infty}^{\infty} P(\mathrm{j}\bar{\omega})\mathrm{e}^{\mathrm{j}\omega t}\mathrm{d}\bar{\omega} \\ P(\mathrm{j}\bar{\omega}) &= \int_{-\infty}^{\infty} p(t)\mathrm{e}^{-\mathrm{j}\bar{\omega}t}\mathrm{d}t \end{aligned} \tag{6.62}$$

where $\bar{\omega}$ represents the loading circular frequency. Analogously, the structural response can be obtained through the frequency domain using the relation

$$w(t) = \frac{1}{2\pi}\int_{-\infty}^{\infty} H(\mathrm{j}\bar{\omega})P(\mathrm{j}\bar{\omega})\mathrm{e}^{\mathrm{j}\bar{\omega}t}\mathrm{d}\bar{\omega} \tag{6.63}$$

in which $H(\mathrm{j}\bar{\omega})$ is the complex frequency response function given by

$$H(\mathrm{j}\bar{\omega}) = \frac{1}{k}\left[\frac{1}{(1-\beta^2)+\mathrm{j}(2\zeta\beta)}\right], \qquad \zeta \geq 0 \tag{6.64}$$

and $\beta \equiv \bar{\omega}/\omega$. Finally, it should be mentioned that the time and frequency domain transfer functions are related through the Fourier transform pair

$$\begin{aligned} H(\mathrm{j}\bar{\omega}) &= \int_{-\infty}^{\infty} h(t)\mathrm{e}^{-\mathrm{j}\bar{\omega}t}\mathrm{d}t \\ h(t) &= \frac{1}{2\pi}\int_{-\infty}^{\infty} H(\mathrm{j}\bar{\omega})\mathrm{e}^{\mathrm{j}\bar{\omega}t}\mathrm{d}\bar{\omega} \end{aligned} \tag{6.65}$$

Remember that the presented technique of response analysis was performed only for an SDOF system.

Furthermore, for structures having many degrees of freedom, MDOF (multi-degree-of-freedom) systems, the so-called mode superposition technique is recommended as a very effective means of evaluating the dynamic structural response. However, the computational cost in this type of calculation is transferred from the MDOF dynamic analysis to the solution of the N degree-of-freedom undamped eigenproblem followed by the modal coordinate transformation,which must be done before the individual modal responses can be evaluated. Here it must be recalled that the equations of motion will be uncoupled by the resulting mode shapes only if the damping is represented by a proportional damping matrix. One approach to the solution of the set of coupled equations

$$\boldsymbol{m}\ddot{\boldsymbol{w}}(t) + \boldsymbol{c}\dot{\boldsymbol{w}}(t) + \boldsymbol{k}\boldsymbol{w}(t) = \boldsymbol{p}(t) \tag{6.66}$$

that may often be worth consideration is the step-by-step procedure, (e.g. the Newmark β, Wilson θ Method).

For linear systems, however, a more convenient solution may be obtained using the Fourier transform (frequency domain) procedures, as well by applying convolution integral (time domain) methods. First is considered the case where the MDOF system is subjected to a unit-impulse loading in the jth degree of freedom, while no other loads are applied. Therefore, the force vector $\boldsymbol{p}(t)$ consists only of zero components except for the jth term. This term can be expressed by $p_j(t) = \delta(t)$, where $\delta(t)$ is the Dirac delta function defined as

$$\delta(t) = \begin{cases} 0, & t \neq 0 \\ \infty, & t = 0 \end{cases} \qquad \int_{-\infty}^{\infty} \delta(t)\mathrm{d}t = 1 \tag{6.67}$$

Assuming now that equation (6.66) can be solved for the displacements caused by this loading, the ith component of the resulting displacement vector $\boldsymbol{w}(t)$ will then be the free-vibration response in that degree of freedom caused by a unit-impulse loading in coordinate j. Thus, by definition this i component motion is a unit-impulse transfer function, which can be denoted by $h_{ij}(t)$. Next, if the loading in coordinate j is assumed to be a general time-varying load $p_j(t)$, the dynamic response in coordinate i could be obtained by superimposing the effects of a succession of impulses in the manner of the Duhamel integral with zero initial conditions. The generalized expression for the response in coordinate i due to the load at j is the following convolution integral:

$$w_{ij}(t) = \int_0^t p_j(\tau)h_{ij}(t-\tau)\mathrm{d}\tau, \qquad i = 1, 2, 3, \ldots, N \tag{6.68}$$

and the total response in coordinate i produced by a general loading involving all components of the load vector $\boldsymbol{p}(t)$ is obtained by summing the contributions from all load components, as follows:

$$w_i(t) = \sum_{j=1}^{N} \left[\int_0^t p_j(\tau)h_{ij}(t-\tau)\mathrm{d}\tau \right], \qquad i = 1, 2, 3, \ldots, N \tag{6.69}$$

For the purpose of the frequency domain analysis it will be assumed that both the load and the response are harmonic. Therefore, the loading is an applied force vector $\boldsymbol{p}(t)$ having all zero components except for the j term, which is a unit harmonic loading, $p_j(t) = 1\exp(\mathrm{j}\bar{\omega}t)$. Note that j in the parentheses stands for the imaginary unit. The resulting steady-state response in the ith component of the displacement vector $w(t)$ can be obtained as

$$w_{ij}(t) = H_{ij}(\mathrm{j}\bar{\omega})\mathrm{e}^{\mathrm{j}\bar{\omega}t} \tag{6.70}$$

where $H_{ij}(\mathrm{j}\bar{\omega})$ is the complex frequency response transfer function, as mentioned above. Superimposing the effects of all the harmonics contained in $p_j(t)$ leads to the total force vibration response, as follows (assuming zero initial conditions):

$$w_{ij}(t) = \frac{1}{2\pi}\int_{-\infty}^{\infty} H_{ij}(\mathrm{j}\bar{\omega})P_j(\mathrm{j}\bar{\omega})\mathrm{e}^{\mathrm{j}\bar{\omega}t}\mathrm{d}\bar{\omega} \tag{6.71}$$

where

$$P_j(\mathrm{j}\bar{\omega}) = \int_{-\infty}^{\infty} p_j(t)\mathrm{e}^{-\mathrm{j}\bar{\omega}t}\mathrm{d}t \tag{6.72}$$

is the Fourier transform of the time domain expression for the loading. Finally, the total response in the ith coordinate produced by a general loading involving all components of the load vector $p(t)$ can be written as

$$w_i(t) = \sum_{j=1}^{N}\left[\frac{1}{2\pi}\int_{-\infty}^{\infty} H_{ij}(\mathrm{j}\bar{\omega})P_{\mathrm{j}}(\mathrm{j}\bar{\omega})\mathrm{e}^{\mathrm{j}\bar{\omega}t}\mathrm{d}\bar{\omega}\right], \qquad i = 1, 2, 3, \ldots, N \tag{6.73}$$

The successful implementation of equations (6.69) and (6.73) depends on being able to generate the transfer functions $h_{ij}(t)$ and $H_{ij}(\mathrm{j}\bar{\omega})$ efficiently. The Fourier transform of the function $w_{ij}(t)$ is the complex function $W_{ij}(\mathrm{j}\bar{\omega})$, defined as follows:

$$W_{ij}(\mathrm{j}\bar{\omega}) = \int_{-\infty}^{\infty}\left[\int_{-\infty}^{t} p_j(\tau)h_{ij}(t-\tau)\mathrm{d}\tau\right]\mathrm{e}^{-\mathrm{j}\bar{\omega}t}\mathrm{d}t \tag{6.74}$$

which can also be written as

$$W_{ij}(\mathrm{j}\bar{\omega}) = H_{ij}(\mathrm{j}\bar{\omega})P_j(\mathrm{j}\bar{\omega}) \tag{6.75}$$

Thus, the interrelationship between both transfer functions is given in a similar way to equation (6.65) by

$$\begin{aligned} H_{ij}(\mathrm{j}\bar{\omega}) &= \int_{-\infty}^{\infty} h_{ij}(t)\mathrm{e}^{-\mathrm{j}\bar{\omega}t}\mathrm{d}t \\ h_{ij}(t) &= \frac{1}{2\pi}\int_{-\infty}^{\infty} H_{ij}(\mathrm{j}\bar{\omega})\mathrm{e}^{\mathrm{j}\bar{\omega}t}\mathrm{d}\bar{\omega} \end{aligned} \tag{6.76}$$

Consider now the Fourier transform of equation (6.66) as expressed in the frequency domain by

$$\left[\left(\boldsymbol{k}-\bar{\omega}^2\boldsymbol{m}\right)+\mathrm{j}(\bar{\omega}\boldsymbol{c})\right]\boldsymbol{W}(\mathrm{j}\bar{\omega})=\boldsymbol{P}(\mathrm{j}\bar{\omega}) \tag{6.77}$$

in which the complex matrix in the bracket term on the left-hand side is the impedance (dynamic stiffness) matrix for the complete structural system being represented. Using the solution procedure presented subsequently, it is not necessary for the viscous damping matrix $\boldsymbol{c}$ to satisfy the orthogonality condition.

Next, equation (6.77) can be written, in abbreviated form, as:

$$\boldsymbol{I}(\mathrm{j}\bar{\omega})\boldsymbol{W}(\mathrm{j}\bar{\omega})=\boldsymbol{P}(\mathrm{j}\bar{\omega}) \tag{6.78}$$

where the impedance matrix $\boldsymbol{I}(\mathrm{j}\bar{\omega})$ is given by the entire bracket matrix on the left-hand side. Premultiplying both sides of this equation by the inverse of the impedance matrix leads to the following expression for the response vector $\boldsymbol{W}(\mathrm{j}\bar{\omega})$:

$$\boldsymbol{W}(\mathrm{j}\bar{\omega})=\boldsymbol{I}(\mathrm{j}\bar{\omega})^{-1}\boldsymbol{P}(\mathrm{j}\bar{\omega}) \tag{6.79}$$

which implies that multiplying a complex matrix by its inverse results in the identity matrix. Although computer programs are readily available to carrying out this type of inversion solution, the approach requires an excessive amount of computational time. This can be reduced by first solving for the complex frequency response transfer functions $H_{ij}(\mathrm{j}\bar{\omega})$ at a set of widely spaced discrete values of $\bar{\omega}$, and then using an effective and efficient interpolation procedure to obtain the transfer functions at the intermediate closely spaced discrete values of $\bar{\omega}$ required by the FFT procedure. By definition the complex frequency response transfer functions are given as

$$\langle H_{1j}(\mathrm{j}\bar{\omega})\; H_{2j}(\mathrm{j}\bar{\omega})\; H_{3j}(\mathrm{j}\bar{\omega})\cdots H_{Nj}(\mathrm{j}\bar{\omega})\rangle^{\mathrm{T}}=\boldsymbol{I}(\mathrm{j}\bar{\omega})^{-1}\boldsymbol{I}_j,\qquad j=1,2,3,\ldots,N \tag{6.80}$$

in which $\boldsymbol{I}_j$ denotes an N-component vector containing all zeros except for the jth component, which equals unity. Because these transfer functions are smooth, even though they peak at the natural frequencies of the system, interpolation procedures can be effectively used in order to obtain their complex values at the intermediate closely spaced discrete values of $\bar{\omega}$.

Finally, using equation (6.80) for the obtained transfer functions, the response vector $\boldsymbol{W}(\mathrm{j}\bar{\omega})$ is given by

$$\boldsymbol{W}(\mathrm{j}\bar{\omega})=\boldsymbol{H}(\mathrm{j}\bar{\omega})\boldsymbol{P}(\mathrm{j}\bar{\omega}) \tag{6.81}$$

in which $\boldsymbol{H}(\mathrm{j}\bar{\omega})$ is the $N \times N$ complex frequency response transfer matrix

$$\boldsymbol{H}(\mathrm{j}\bar{\omega}) = \begin{bmatrix} H_{11}(\mathrm{j}\bar{\omega}) & H_{12}(\mathrm{j}\bar{\omega}) & \dots & H_{1N}(\mathrm{j}\bar{\omega}) \\ H_{21}(\mathrm{j}\bar{\omega}) & H_{22}(\mathrm{j}\bar{\omega}) & \dots & H_{2N}(\mathrm{j}\bar{\omega}) \\ \vdots & \vdots & \vdots & \vdots \\ H_{N1}(\mathrm{j}\bar{\omega}) & H_{N2}(\mathrm{j}\bar{\omega}) & \dots & H_{NN}(\mathrm{j}\bar{\omega}) \end{bmatrix} \tag{6.82}$$

obtained for each frequency required in the response analysis. Note that once this transfer matrix has been obtained, the responses of the system to multiple sets of loadings can be obtained by Fourier transforming and then multiplying the resulting vector set in each case by the transfer matrix, in accordance with equation (6.81). Having the vector $\boldsymbol{W}(\mathrm{j}\bar{\omega})$ for each set, the corresponding set of displacements in vector $\boldsymbol{w}(t)$ can be obtained by means of the inverse Fourier transformation.

The interpolation procedure for generation of transfer functions will be shown by means of an example. Another transfer function technique for the solution of the equation of motion, equation (6.66), will be illustrated, namely the Laplace transformation. First, an SDOF system should be considered, where its equation of motion can be expressed as

$$m\ddot{w}(t) + c\dot{w}(t) + kw(t) = p(t) \tag{6.83}$$

In general, this is a second-order differential equation with initial conditions. Thus, taking the Laplace transform gives

$$\mathcal{L}\{\ddot{w}(t)\} = s^2 W(s) - sW(0) - \dot{W}(0) \tag{6.84}$$

where $W(0)$ and $\dot{W}(0)$ are the displacement and velocity initial conditions respectively and $W(s)$ is the Laplace transform of $W(t)$. Assuming now zero initial conditions leads to

$$\mathcal{L}\{\ddot{w}(t)\} = s^2 W(s) \tag{6.85}$$

Thus, the Laplace transform of the SDOF equation of motion, equation (6.83), can be written as

$$ms^2 W(s) + csW(s) + kW(s) = P(s) \tag{6.86}$$

where $P(s)$ represents the Laplace transform of the loading $p(t)$. Solving for the transfer function gives

$$\frac{W(s)}{P(s)} = \frac{1}{ms^2 + cs + k} = \frac{\frac{1}{m}}{s^2 + \frac{c}{m}s + \frac{k}{m}} \tag{6.87}$$

Next, this equation will be simplified by means of the following definitions:

$$\omega_{\mathrm{n}}^2 = \frac{k}{m} \quad \text{undamped natural frequency}$$

$$c_{\mathrm{cr}} = 2\sqrt{km} \quad \text{critical damping value}$$

to an expression in the form

$$\frac{W(s)}{P(s)} = \frac{\frac{1}{m}}{s^2 + 2\zeta\omega_n s + \omega_{\mathrm{n}}^2} \tag{6.88}$$

Substituting by $s \rightarrow \mathrm{j}\bar{\omega}$ allows the frequency response to be calculated:

$$\frac{W(\mathrm{j}\bar{\omega})}{P(\mathrm{j}\bar{\omega})} = \frac{\frac{1}{m}}{(\mathrm{j}\bar{\omega})^2 + 2\zeta\omega_{\mathrm{n}}(\mathrm{j}\bar{\omega}) + \omega_{\mathrm{n}}^2} = \frac{\frac{1}{m\bar{\omega}^2}}{\left[\left(\frac{\omega_{\mathrm{n}}^2}{\bar{\omega}}\right) - 1\right] + \mathrm{j}2\zeta\left(\frac{\omega_{\mathrm{n}}}{\bar{\omega}}\right)} \tag{6.89}$$

This frequency response equation shows how the ratio W/P varies as a function of the frequency ω. The ratio is a complex number and has some interesting properties at different values of the ratio $\omega_{\mathrm{n}}/\bar{\omega}$. Its magnitude and phase change respectively are plotted in Figures 6.27 and 6.28.

For MDOF systems, however, the Laplace transform procedure can be used to obtain the system eigenvalues and zeros. Remember that the eigenvalues represent the resonant frequencies. In other words, they show the frequencies

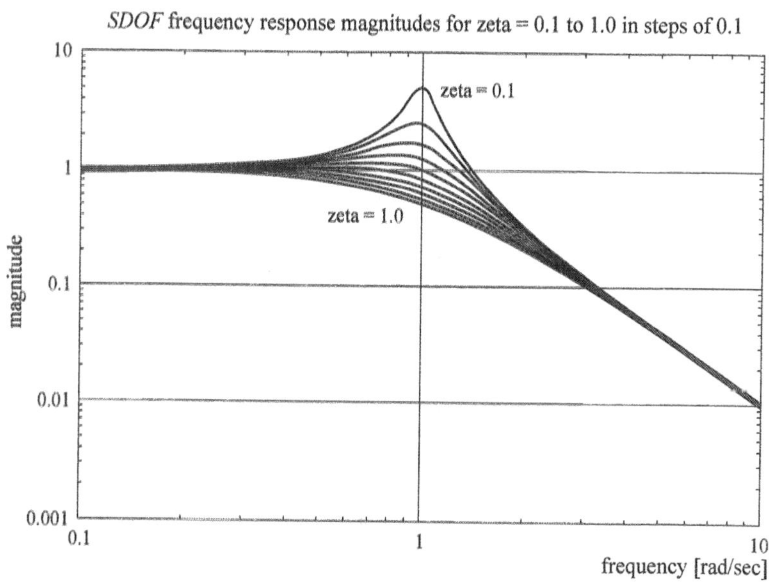

Figure 6.27 SDOF magnitude versus frequency for different damping ratios

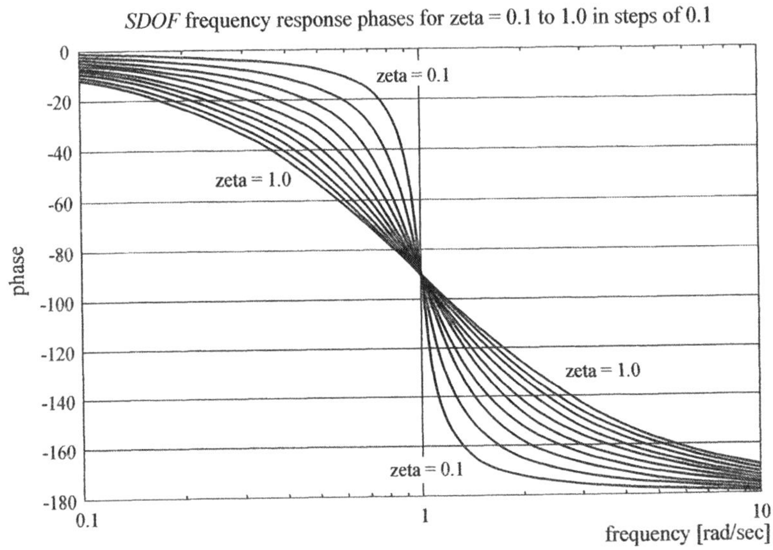

Figure 6.28 SDOF phase versus frequency for different damping ratios

where the system will amplify inputs. Thus, the eigenvalues are a basic system characteristic that depends only on the distribution of mass, stiffness and damping throughout the system, and not on where the forces are applied or where displacements are measured. On the other hand, the zeros show the frequencies where the system will attenuate inputs. The calculation procedure for an MDOF system occurs as follows:

- Set up the equation of motion similar to equation (6.83).
- Take Laplace transforms assuming zero initial conditions as in equation (6.85).
- Solve for the transform function analogue to equation (6.87).
- Determine the characteristic equation and solve for eigenvalues.
- Finally, determine the zeros for each transfer function.

6.7.3 APPLICATIONS (EXAMPLES)

6.7.3.1 Measurement of Transfer Functions and Prediction of Structural Response using Transfer Functions

The simplest system should be considered in this example, namely an SDOF system. Remember that its equation of motion is

$$m\ddot{w}(t) + c\dot{w}(t) + kw(t) = p(t) \tag{6.90}$$

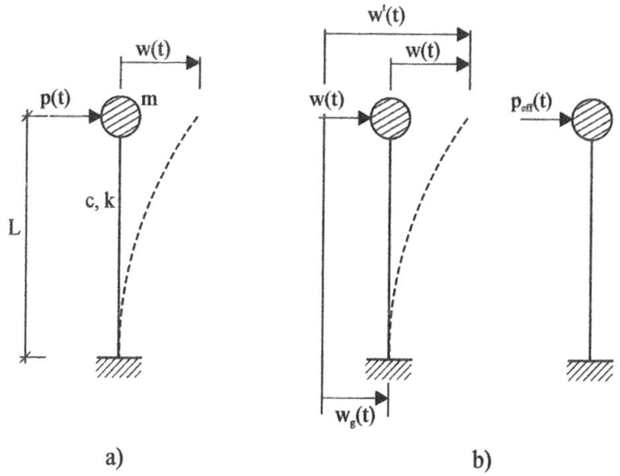

Figure 6.29 SDOF system heavily loaded by (a) the time-varying force; (b) support excitation

where the equation members are shown in Figure 6.29(a). However, the equation of motion due to support excitation (Figure 6.29(b)) can be written as

$$m\ddot{w}(t) + c\dot{w}(t) + kw(t) = -m\ddot{w}_g(t) \equiv p_{\text{eff}}(t) \tag{6.91}$$

where $p_{\text{eff}}(t)$ denotes the effective support excitation loading. In other words, the structural deformations caused by ground acceleration $\ddot{w}_g(t)$ are exactly the same as those produced by an external load $p(t) = -m\ddot{w}_g(t)$. Note that the negative sign shows that the effective force opposes the sense of ground acceleration.

For the numerical analyses presented below the following structural parameters are assumed:

- Mass $m = 40\ 035$ kg.
- Damping ratio $\zeta = 0.05$.
- Stiffness $k = 11\ 398.557$ N/m.

The system response due to a 20 Hz band-limited white noise support excitation (Figure 6.30) is analysed. The system response through the time domain (Figure 6.31) is obtained by means of the time-integration technique, namely the Newmark method.

The transfer function can be obtained theoretically as well as experimentally. First the theoretical transfer function is developed using equation (6.87). Substituting the expression for an effective support excitation load in this equation leads to

$$H(\mathrm{j}\bar{\omega}) = \frac{W(\mathrm{j}\bar{\omega})}{\ddot{W}_g(\mathrm{j}\bar{\omega})} = -\frac{m}{m(\mathrm{j}\bar{\omega})^2 + c(\mathrm{j}\bar{\omega}) + k} = -\frac{1}{(\mathrm{j}\bar{\omega})^2 + \frac{c}{m}(\mathrm{j}\bar{\omega}) + \frac{k}{m}} \tag{6.92}$$

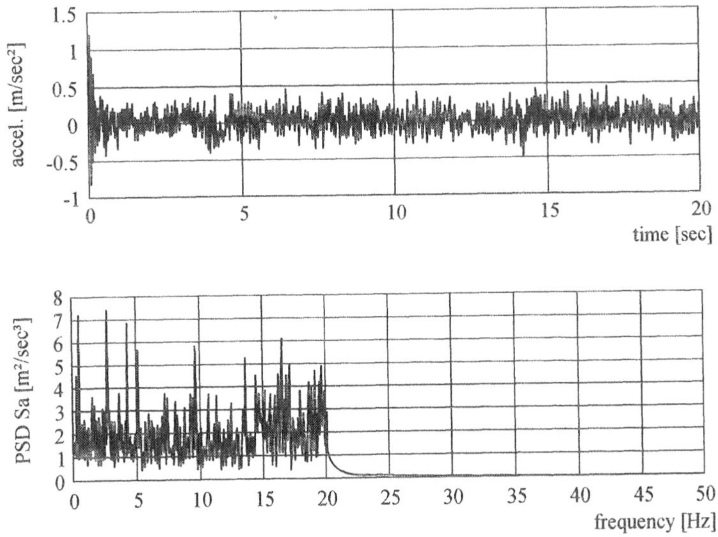

Figure 6.30 White noise ground motion: time history and power spectral density (PSD) plots respectively

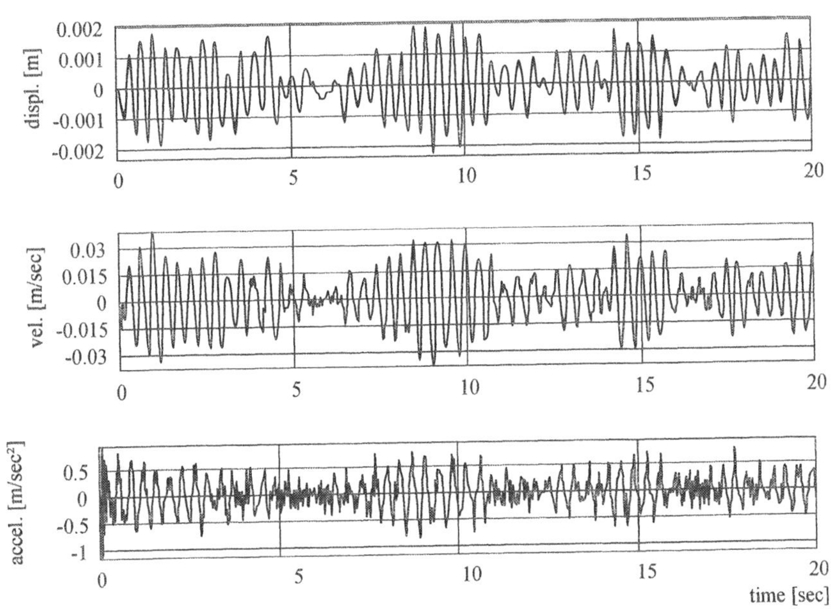

Figure 6.31 System response through the time domain

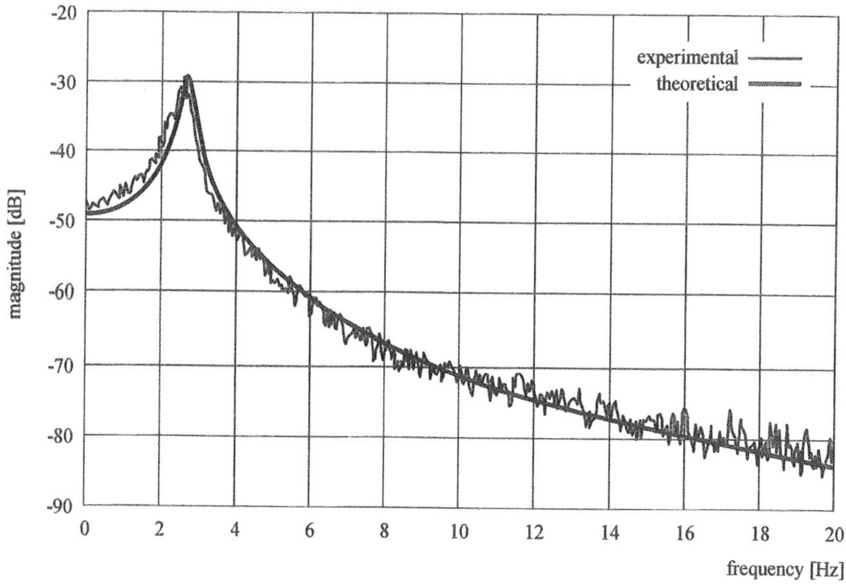

Figure 6.32 SDOF transfer function obtained theoretically and experimentally

Next, assume that the system response, $w(t)$, can be measured and the support excitation, $\ddot{w}_g(t)$, is known. Thus, the experimental transfer function can be calculated by the relationship

$$H(j\bar{\omega}) = \frac{W(j\bar{\omega})}{\ddot{W}_g(j\bar{\omega})} \tag{6.93}$$

Figure 6.32 shows the obtained transfer functions, where a good match can be realized. Assuming that the theoretical transfer function and the support excitation are known, a prediction of the structural response (Figure 6.33) can be calculated. For this purpose equation (6.93) is rewritten as

$$W(j\bar{\omega}) = H(j\bar{\omega})\ddot{W}_g(j\bar{\omega}) \tag{6.94}$$

6.7.3.2 Damage Detection by Means of Transfer Functions [12]

This example shows the application of so-called Component Transfer Functions in order to detect failure source in structures. Interest in practising structural health monitoring (SHM) and then detecting damage at the earliest

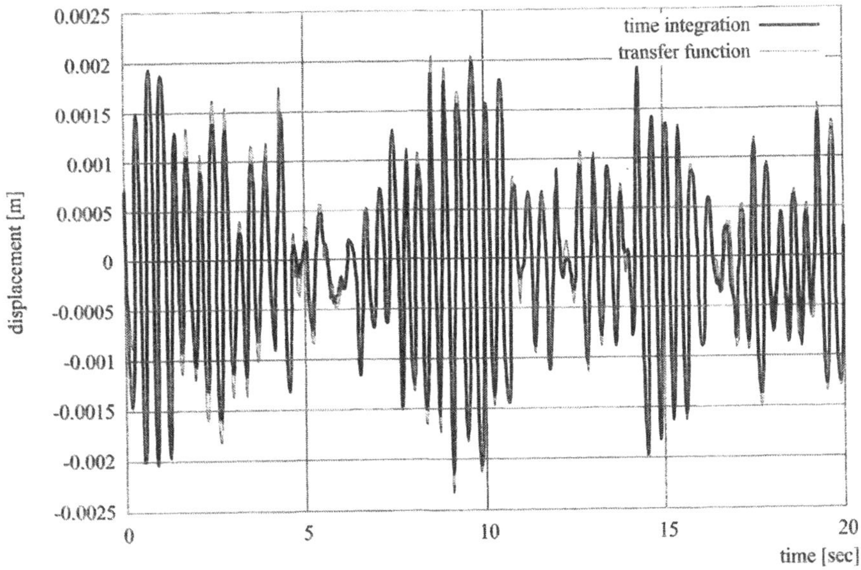

Figure 6.33 SDOF response by time integration and transfer function approach

possible stage has increased throughout the civil engineering community in the last decade. Four levels of damage identification are known to date:

1. Is the structure damaged?
2. Where is the damage located?
3. What is the damage extent?
4. What is the residual structural serviceability?

In general, damage can be classified as linear or nonlinear. Linear damage is observed in the case when an initially linear elastic system remains linear elastic after the occurrence of damage, whereas if the structure behaves as an inelastic way nonlinear damage can be determined.

A three-storey 'shear resisting' steel frame is modelled by FE to study the effectiveness of the proposed system identification (SI) techniques. A set of twenty 20 Hz band-limited white noise signals is generated and each of them is applied as acceleration support excitation, while the acceleration time history at each floor is posted. In order to affect nonlinear damage effects, a plastic hinge is simulated immediately below the first floor by means of cross-section reduction. Figure 6.34 depicts the structural model and a typical white noise signal.

Remember that the transfer function can be represented as the relationship between the input and the output of a system: $H_{ij}(j\bar{\omega}) = W_{ij}(j\bar{\omega})/P_j(j\bar{\omega})$.

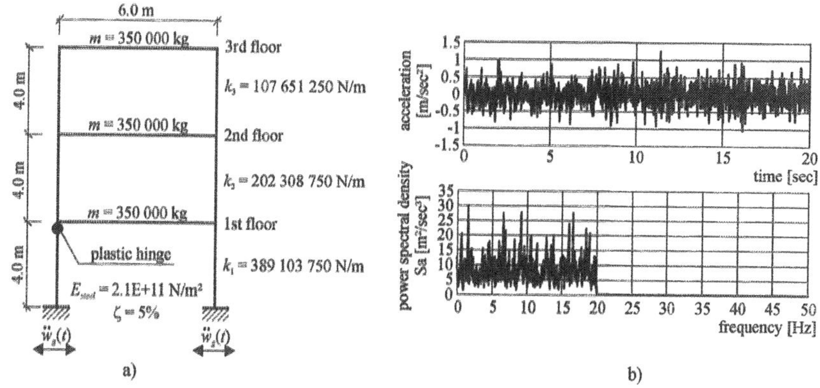

Figure 6.34 System of consideration: (a) structural model; (b) typical 20 Hz band-limited white noise support excitation and its PSD

Consequently, the component transfer function (CTF) can be defined as the relationship between any two outputs of an MDOF system used to determine whether damage is presented in the structure and to identify the location of the damage:

$$\hat{H}_{\nu u}(\mathrm{j}\bar{\omega}) = \frac{W_{\nu j}(\mathrm{j}\bar{\omega})}{W_{uj}(\mathrm{j}\bar{\omega})} \tag{6.95}$$

In order to explain the basic idea of this method, consider a three-storey seismically excited structure as shown in Figure 6.35. A lumping idealization of the structure leads to the following equation of motion:

$$\begin{bmatrix} m_1 & 0 & 0 \\ 0 & m_2 & 0 \\ 0 & 0 & m_3 \end{bmatrix} \begin{Bmatrix} \ddot{w}_1(t) \\ \ddot{w}_2(t) \\ \ddot{w}_3(t) \end{Bmatrix} + \begin{bmatrix} c_1 + c_2 & -c_2 & 0 \\ -c_2 & c_2 + c_3 & -c_3 \\ 0 & -c_3 & c_3 \end{bmatrix} \begin{Bmatrix} \dot{w}_1(t) \\ \dot{w}_2(t) \\ \dot{w}_3(t) \end{Bmatrix}$$

$$+ \begin{bmatrix} k_1 + k_2 & -k_2 & 0 \\ -k_2 & k_2 + k_3 & -k_3 \\ 0 & -k_3 & k_3 \end{bmatrix} \begin{Bmatrix} w_1(t) \\ w_2(t) \\ w_3(t) \end{Bmatrix} = - \begin{Bmatrix} m_1 \\ m_2 \\ m_3 \end{Bmatrix} \ddot{w}_g(t) \tag{6.96}$$

Taking the Laplace transform

$$\mathcal{L}\{\ddot{w}(t)\} = s^2 W(s) - sW(0) - \dot{W}(0) \tag{6.97}$$

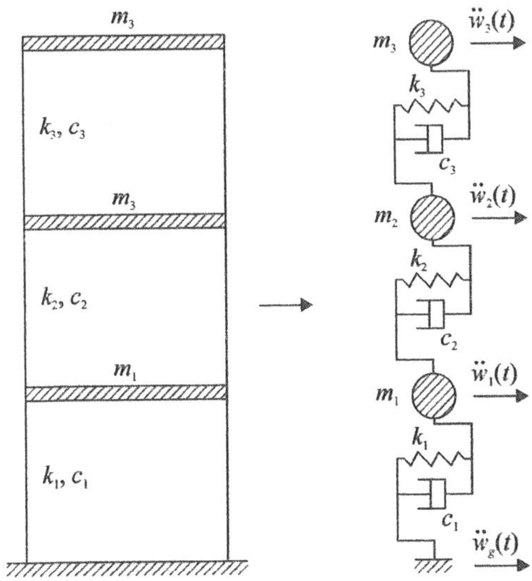

Figure 6.35 Structure of the theoretical consideration

and assuming 'zero' initial conditions, i.e. $W(0) = 0$ and $\dot{W}(0) = 0$, gives the following representation of equation (6.96) through the space domain:

$$\begin{bmatrix} m_1 & 0 & 0 \\ 0 & m_2 & 0 \\ 0 & 0 & m_3 \end{bmatrix} \begin{Bmatrix} s^2 W_1(s) \\ s^2 W_2(s) \\ s^2 W_3(s) \end{Bmatrix} + \begin{bmatrix} c_1 + c_2 & -c_2 & 0 \\ -c_2 & c_2 + c_3 & -c_3 \\ 0 & -c_3 & c_3 \end{bmatrix} \begin{Bmatrix} sW_1(s) \\ sW_2(s) \\ sW_3(s) \end{Bmatrix}$$
$$+ \begin{bmatrix} k_1 + k_2 & -k_2 & 0 \\ -k_2 & k_2 + k_3 & -k_3 \\ 0 & -k_3 & k_3 \end{bmatrix} \begin{Bmatrix} W_1(s) \\ W_2(s) \\ W_3(s) \end{Bmatrix} = -\begin{Bmatrix} m_1 \\ m_2 \\ m_3 \end{Bmatrix} \ddot{W}_g(s) \qquad (6.98)$$

From rearrangement of equation (6.98)

$$\begin{bmatrix} m_1 s^2 + (c_1 + c_2)s + k_1 + k_2 & -c_2 s - k_2 & 0 \\ -c_2 s - k_2 & m_2 s^2 + (c_2 + c_3)s + k_2 + k_3 & -c_3 s - k_3 \\ 0 & -c_3 s - k_3 & m_3 s^2 + c_3 s + k_3 \end{bmatrix}$$
$$\begin{Bmatrix} W_1(s) \\ W_2(s) \\ W_3(s) \end{Bmatrix} = -\begin{Bmatrix} m_1 \\ m_2 \\ m_3 \end{Bmatrix} \ddot{W}_g(s) \qquad (6.99)$$

the CTF can be obtained as the relationships $W_1(s)/\ddot{W}_g(s)$, $W_2(s)/\ddot{W}_g(s)$, $W_3(s)/\ddot{W}_g(s)$, $W_2(s)/W_1(s)$, $W_3(s)/W_2(s)$ and $W_3(s)/W_1(s)$.

In practice, however, the structural response is more often presented by acceleration. Thus, the relationship

$$\frac{\ddot{W}_i(s)}{\ddot{W}_g(s)} = -\frac{s^2 W_i(s)}{\ddot{W}_g(s)} + 1 \tag{6.100}$$

gives the final expressions for the CTFs. Next, damage effects in the columns of the first floor can be simulated by reducing the stiffness of the first floor by 60%. Additionally increasing damping can be assumed as well, e.g. $\zeta_1 = 5\%$, $\zeta_2 = 2\%$ and $\zeta_3 = 0\%$. Figure 6.36 compares the obtained CTFs of the undamaged system with those of the damaged structure. If the peaks of these transfer functions, with regard to the actual system state, shift in comparison to those of a reference system state, structural change is observed. In particular, if the peak frequency values decrease, and assuming the mass is kept constant, a loss of stiffness occurs. In other words, linear damage effects are detected. In addition, a decrease in the peak relative amplitude values is caused by an increase in structural damping, which characterizes the presence of nonlinear damage.

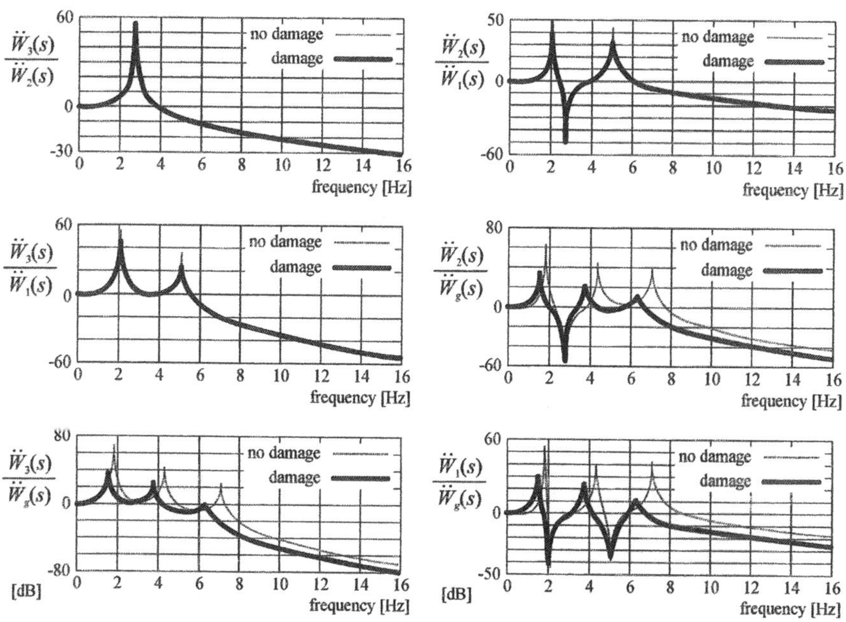

Figure 6.36 Sample CTFs of the undamaged and damaged structures respectively

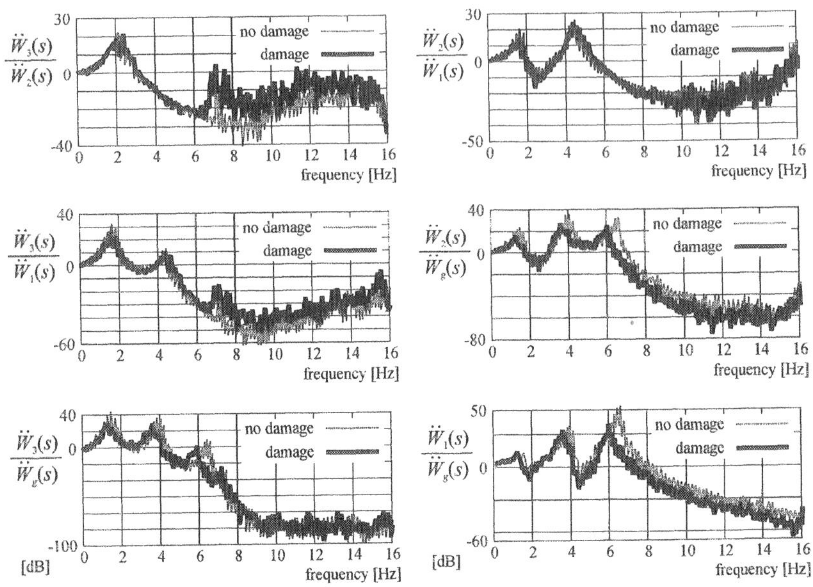

Figure 6.37 Experimentally obtained CTFs of the undamaged and damaged systems respectively

Using equation (6.95) the CTFs are calculated by means of the averaged PSD of the input ground motion and of the corresponding individual floor outputs. The obtained results are shown in Figure 6.37. Note that in $\ddot{W}_3(s)/\ddot{W}_2(s)$, $\ddot{W}_3(s)/\ddot{W}_1(s)$ and $\ddot{W}_2(s)/\ddot{W}_1(s)$ there is no observable loss of stiffness, which means that no damage occurs between the third and first floors. However, nonlinear damage effects are detected in $\ddot{W}_3(s)/\ddot{W}_g(s)$, $\ddot{W}s_2(s)/\ddot{W}_g(s)$ and $\ddot{W}_1(s)/\ddot{W}_g(s)$. In other words, the damage is located between the first and the base floors. Therefore, level 2 of damage detection can be provided by this technique.

6.8 STOCHASTIC SUBSPACE IDENTIFICATION

Contributed by Guido De Roeck, Bart Peeters and Anne Teughels, K.U. Leuven, Belgium

Several models of vibrating structures exist, going from models that are close to physical reality towards general dynamic models that are useful in system identification. Examples of these model types are FE models of civil engineering structures, state-space models originating from electrical engineering and modal models initially developed in mechanical engineering.

System identification starts by adopting a certain model that is believed to represent the system. Next, values are assigned to the parameters of the model so as to match the measurements. Stochastic system identification methods estimate the parameters of stochastic models by using output-only data. The methods can be divided according to the type of data that they require: frequency domain spectral data, covariances or raw time data. Accordingly, they evolve from picking the peaks of spectral densities to subspace methods that make extensive use of concepts from numerical linear algebra.

In a civil engineering context, the civil structures (e.g. bridges, towers) are the systems; estimation of the modal parameters is a particular type of identification and stochastic means that the structure is excited by a not-measurable input force and that only output measurements (e.g. acceleration) are available. It is assumed that the input corresponds to white noise.

The time domain data-driven stochastic methods identify models directly from the response time signals. The data-driven stochastic subspace identification (SSI) method cancels out the (uncorrelated) noise by projecting the row space of future outputs into the row space of past outputs. The idea behind this projection is that it retains all the information in the past that is useful to predict the future. Robust numerical techniques from linear algebra, such as QR factorization (an algorithm for matrix decomposition into a Q part and an R part), singular value decomposition and least squares, are used in the future processing of the data in order to solve the identification problem. The principles of a non-steady-state Kalman filter are applied for the identification of a state-space model.

Once the parametric model is identified and available, the modal parameters can then easily be derived from the model matrices. In practice, civil structures are frequently excited by ambient forces (such as wind, traffic) or impact loads (coming from a hammer or a drop weight). The main advantage of ambient sources is the fact that the bridges can stay operational, which avoids the costs that would evolve from putting them out of use. On the contrary, artificial excitation by a shaker is not very cost-effective, since a very powerful shaker is necessary to excite the heavy structure and additional manpower is needed to install it. Furthermore, if a structure has low-frequency (below 1 Hz) modes, it may be difficult to excite it with a shaker, whereas this is generally no problem for a drop weight or ambient sources. The high-frequency modes, on the other hand, are not always well excited by ambient sources. If mass-normalized mode shapes are required, ambient excitation cannot be used. To obtain the correct scaling of the mode shapes, the applied force has to be known.

6.8.1 STOCHASTIC STATE-SPACE MODELS

6.8.1.1 The Stochastic Components

The dynamic behaviour of a mechanical system, discretized by finite elements, is described by the matrix differential equation

$$\boldsymbol{M}\ddot{\boldsymbol{U}}(t) + \boldsymbol{C}\dot{\boldsymbol{U}}(t) + \boldsymbol{K}\boldsymbol{U}(t) = \boldsymbol{R}(t) \tag{6.101}$$

where $\boldsymbol{M}$, $\boldsymbol{C}$ and $\boldsymbol{K}$ are the mass, damping and stiffness matrices, $\boldsymbol{U}(t)$ is the displacement vector at continuous time t and $\boldsymbol{R}(t)$ is the excitation vector. As shown in reference [13], this description of the dynamic behaviour can be converted in a discrete-time state-space model:

$$\begin{aligned} \boldsymbol{x}_{k+1} &= \boldsymbol{A}\boldsymbol{x}_k + \boldsymbol{B}\boldsymbol{u}_k \\ \boldsymbol{y}_k &= \boldsymbol{C}\boldsymbol{x}_k + \boldsymbol{D}\boldsymbol{u}_k \end{aligned} \tag{6.102}$$

where $\boldsymbol{x}_k = \boldsymbol{x}(k\,\Delta t)$ is the discrete-time state vector containing the sampled displacements and velocities and $\boldsymbol{u}_k$, $\boldsymbol{y}_k$ are the sampled input (measured force) and output (measured displacements, velocities or accelerations). The matrices $\boldsymbol{A}$, $\boldsymbol{B}$, $\boldsymbol{C}$ and $\boldsymbol{D}$ can be related to the original matrices $\boldsymbol{M}$, $\boldsymbol{C}$ and $\boldsymbol{K}$ [14].

A final step towards the experimental world means adding of noise. Up to now it was assumed that the system was only driven by a deterministic input $\boldsymbol{u}_k$. However, the deterministic models are not able to describe real measurement data exactly. Stochastic components have to be included in the models and the following discrete-time combined deterministic-stochastic state-space model is obtained:

$$\begin{aligned} \boldsymbol{x}_{k+1} &= \boldsymbol{A}\boldsymbol{x}_k + \boldsymbol{B}\boldsymbol{u}_k + \boldsymbol{w}_k \\ \boldsymbol{y}_k &= \boldsymbol{C}\boldsymbol{x}_k + \boldsymbol{D}\boldsymbol{u}_k + \boldsymbol{v}_k \end{aligned} \tag{6.103}$$

where $\boldsymbol{w}_k$ is the process noise due to disturbances and modelling inaccuracies and $\boldsymbol{v}_k \in \mathbb{R}^l$ is the measurement noise due to sensor inaccuracy. They are both not-measurable vector signals assumed to be zero mean, white and with covariance matrices:

$$\boldsymbol{E}\left[\begin{pmatrix} w_p \\ v_p \end{pmatrix} \begin{pmatrix} w_q^{\mathrm{T}} & v_q^{\mathrm{T}} \end{pmatrix}\right] = \begin{pmatrix} Q & S \\ S^{\mathrm{T}} & R \end{pmatrix} \delta_{pq} \tag{6.104}$$

where $\boldsymbol{E}$ is the expected value operator, δ_{pq} is the Kronecker delta (if $p = q$ then $\delta_{pq} = 1$, otherwise $\delta_{pq} = 0$) and p, q are two arbitrary time instants.

In a civil engineering context, the only vibration information that is available are the responses of a structure excited by some not-measurable inputs. Due to the lack of input information it is not possible (from a system identification point of view) to distinguish between the terms in $\boldsymbol{u}_k$ and the noise terms $\boldsymbol{w}_k$ and $\boldsymbol{v}_k$. The discrete-time stochastic state-space model finally reads

$$\begin{aligned} \boldsymbol{x}_{k+1} &= \boldsymbol{A}\boldsymbol{x}_k + \boldsymbol{w}_k \\ \boldsymbol{y}_k &= \boldsymbol{C}\boldsymbol{x}_k + \boldsymbol{v}_k \end{aligned} \tag{6.105}$$

The input is now implicitly modelled by the noise terms. However, the white noise assumptions of these terms cannot be omitted: it is necessary for the proofs of the system identification methods of the next section. The consequence is that if this white noise assumption is violated, for instance if the input also contains some dominant frequency components in addition to white noise, these frequency components cannot be separated from the eigenfrequencies of the system and will appear as (spurious) poles of the state matrix A.

6.8.1.2 Properties of Stochastic Systems

Some important properties of stochastic systems are briefly resumed. They are well known and can, for instance, be found in reference [15]. As already stated, the noise terms have zero mean and their covariance matrices are given later by equation (6.111). There are some further assumptions. The stochastic process is assumed to be stationary with zero mean:

$$\boldsymbol{E}[x_k\, x_k^{\mathrm{T}}] = \boldsymbol{\Sigma}, \qquad \boldsymbol{E}[x_k] = 0 \tag{6.106}$$

where the state covariance matrix Σ is independent of the time k. Since w_k, v_k have zero mean and are independent of the actual state,

$$\boldsymbol{E}[x_k\, w_k^{\mathrm{T}}] = \boldsymbol{\Sigma}, \qquad \boldsymbol{E}[x_k\, \nu_k^{\mathrm{T}}] = 0 \tag{6.107}$$

The output covariance matrices $\boldsymbol{R}_i \in \mathbb{R}^{l\times l}$ are defined as

$$\boldsymbol{R}_i = \boldsymbol{E}[y_{k+i}\, y_k^{\mathrm{T}}] \tag{6.108}$$

where i is an arbitrary time lag. The 'next state output' covariance matrix $\boldsymbol{G} \in \mathbb{R}^{n\times l}$ is defined as

$$\boldsymbol{G} = \boldsymbol{E}[x_{k+1}\, y_k^{\mathrm{T}}] \tag{6.109}$$

From stationarity, the noise properties and previous definitions following properties are easily deduced:

$$\begin{aligned} \boldsymbol{\Sigma} &= \boldsymbol{A\Sigma A}^{\mathrm{T}} + \boldsymbol{Q} \\ \boldsymbol{R}_0 &= \boldsymbol{C\Sigma C}^{\mathrm{T}} + \boldsymbol{R} \\ \boldsymbol{G} &= \boldsymbol{A\Sigma C}^{\mathrm{T}} + \boldsymbol{S} \end{aligned} \tag{6.110}$$

and for $i = 1, 2, \ldots,$

$$\begin{aligned} \boldsymbol{R}_i &= \boldsymbol{CA}^{i-1}\boldsymbol{G} \\ \boldsymbol{R}_{-i} &= \boldsymbol{G}^{\mathrm{T}}(\boldsymbol{A}_{i-1})^{\mathrm{T}}\boldsymbol{C}^{\mathrm{T}} \end{aligned} \tag{6.111}$$

The last property is very important. This equation alone nearly constitutes the solution to the identification problem: the output covariance sequence can be estimated from the measurement data. Therefore if it is possible to decompose the estimated output covariance sequence according to the last equation of (6.111), the state-space matrices are found.

6.8.1.3 The Forward Innovation Model

An alternative model for stochastic systems that is more suitable for some applications is the so-called forward innovation model. It is obtained by applying the steady-state Kalman filter to the stochastic state-space model equation (6.105):

$$\begin{aligned} z_{k+1} &= \boldsymbol{A} z_k + \boldsymbol{K} \boldsymbol{e}_k \\ y_k &= \boldsymbol{C} z_k + \boldsymbol{e}_k \end{aligned} \tag{6.112}$$

The elements of the sequence $\boldsymbol{e}_k$ are called innovations – hence the name of the model. It is a white noise vector sequence, with the covariance matrix

$$\boldsymbol{E}[e_p e_q^{\mathrm{T}}] = \boldsymbol{R}_{\mathrm{e}} \delta_{\mathrm{pq}} \tag{6.113}$$

The computation of the forward innovation model ($\boldsymbol{A}$, $\boldsymbol{K}$, $\boldsymbol{C}$, $\boldsymbol{R}_e$) from the stochastic state-space model ($\boldsymbol{A}$, $\boldsymbol{G}$, $\boldsymbol{C}$, $\boldsymbol{R}_0$) starts by finding the positive definite solution $\boldsymbol{P}$ of the discrete Riccati equation:

$$\boldsymbol{P} = \boldsymbol{A}\boldsymbol{P}\boldsymbol{A}^{\mathrm{T}} + (\boldsymbol{G} - \boldsymbol{A}\boldsymbol{P}\boldsymbol{C}^{\mathrm{T}})(\boldsymbol{R}_0 - \boldsymbol{C}\boldsymbol{P}\boldsymbol{C}^{\mathrm{T}})^{-1}(\boldsymbol{G} - \mathbf{APC}^{\mathrm{T}})^{\mathrm{T}} \tag{6.114}$$

The matrix $\boldsymbol{P} \in \mathbb{R}^{n \times n}$ is the forward state covariance matrix $\boldsymbol{P} = \mathbf{E}[z_k\, z_k^{\mathrm{T}}]$. The Kalman gain is then computed as

$$\boldsymbol{K} = (\boldsymbol{G} - \boldsymbol{A}\boldsymbol{P}\boldsymbol{C}^{\mathrm{T}})(\boldsymbol{R}_0 - \boldsymbol{C}\boldsymbol{P}\boldsymbol{C}^{\mathrm{T}})^{-1} \tag{6.115}$$

and the covariance matrix of the innovations equals

$$\boldsymbol{R}_e = \boldsymbol{R}_0 - \boldsymbol{C}\boldsymbol{R}\boldsymbol{C}^{\mathrm{T}} \tag{6.116}$$

6.8.2 STOCHASTIC SYSTEM IDENTIFICATION

This section deals with stochastic system identification methods. In a civil engineering context, structures such as bridges and towers are the systems; the estimation of the modal parameters is the particular type of identification and stochastic means that the structure is excited by an unmeasurable input force and that only output measurements (e.g. accelerations) are available. In these methods the deterministic knowledge of the input is replaced by the assumption that the input is a realization of a stochastic process (white noise).

In this section the time-domain data-driven stochastic subspace identification is described. Other variants (frequency domain spectrum-driven) exist (see reference [14]).

6.8.2.1 Time Data

In principle (output) data y_k are available as discrete samples of the time signal. Measurements for modal analysis applications typically contain some redundancy. Since the spatial resolution of the experimental mode shapes is determined by the position and the number of the sensors, usually many sensors (mostly accelerometers) are used in a modal analysis experiment. Theoretically, if none of the sensors is placed at a node of a mode, all signals carry the same information on eigenfrequencies and damping ratios. To decrease this redundancy, some signals are partially omitted in the identification process, leading to algorithms that are faster and require less computer memory without losing a lot of accuracy. In the end, the omitted sensors are again included to yield the 'full' mode shapes. Assume that the l outputs are split in a subset of r well-chosen reference sensors and a subset of $l-r$ other sensors, and that they are arranged so as to have the references first:

$$\boldsymbol{y}_k = \begin{pmatrix} y_k^{\text{ref}} \\ y_k^{\sim\text{ref}} \end{pmatrix}, \qquad y_k^{\text{ref}} = \boldsymbol{L}\boldsymbol{y}_k, \qquad \boldsymbol{L} = (I_r\, 0) \tag{6.117}$$

where $y_k^{ref} \in \mathbb{R}^r$ are the reference outputs and $y_k^{-ref} \in \mathbb{R}^{l-r}$ are the others; $\boldsymbol{L} \in \mathbb{R}^{r \times l}$ is the selection matrix that selects the references. The choice of the reference sensors in output-only modal analysis corresponds to the choice of the input locations in traditional input–output modal analysis [16, 17].

It is useful in the development of some of the identification methods to gather the output measurements in a block Hankel matrix with $2i$ block rows and N columns. The first i blocks have r rows; the last i have l rows. For the statistical proves of the methods, it is assumed that $N \to \infty$. The Hankel matrix $\boldsymbol{H}^{\text{ref}} \in \mathbb{R}^{(r+1)i \times N}$ can be divided into a past reference and a future part:

$$\boldsymbol{H}^{\text{ref}} = \frac{1}{N^{\frac{1}{2}}} \left(\begin{array}{cccc} y_0^{\text{ref}} & y_1^{\text{ref}} & \cdots & y_{N-1}^{\text{ref}} \\ y_1^{\text{ref}} & y_2^{\text{ref}} & \cdots & y_N^{\text{ref}} \\ \vdots & \vdots & \ddots & \vdots \\ y_{i-1}^{\text{ref}} & y_i^{\text{ref}} & \cdots & y_{i+N-2}^{ref} \\ \hline y_i & y_{i+1} & \cdots & y_{i+N-1} \\ y_{i+1} & y_{i+2} & \cdots & y_{i+N} \\ \vdots & \vdots & \ddots & \vdots \\ y_{2i} - 1 & y_{2i} & \cdots & y_{2i+N-2} \end{array} \right) = \left(\begin{array}{c} \boldsymbol{Y}_{0|i-1}^{\text{ref}} \\ \hline \boldsymbol{Y}_{i|2i-1} \end{array} \right) = \left(\begin{array}{c} \boldsymbol{Y}_{\text{p}}^{\text{ref}} \\ \hline \boldsymbol{Y}_{\text{f}} \end{array} \right) \begin{array}{l} \updownarrow ri \text{ 'past'} \\ \updownarrow li \text{ 'future'} \end{array} \tag{6.118}$$

Note that the output data is scaled by a factor $1/N^{\frac{1}{2}}$. The subscripts of $\boldsymbol{Y}_{i/2i-1} \in \mathbb{R}^{1i \times N}$ are the subscripts of the first and last element in the first column of the block Hankel matrix. The subscripts p and f stand for past and future.

The matrices $Y_{\mathrm{p}}^{\mathrm{ref}}$ and Y_{f} are defined by splitting H^{ref} into two parts of i block rows. Another division is obtained by adding one block row to the past references and omitting the first block row of the future outputs. Because the references are only a subset of the outputs, $l-r$ rows are left over in this new division. These rows are denoted by $Y_{i|i}^{\sim\mathrm{ref}} \in \mathbb{R}^{(l-r)\times N}$:

$$H^{\mathrm{ref}} = \left(\begin{array}{c} Y_{0|i}^{\mathrm{ref}} \\ \hline Y_{i|i}^{\sim\mathrm{ref}} \\ \hline Y_{i+1|2i-1} \end{array}\right) = \left(\begin{array}{c} Y_{\mathrm{p}}^{\mathrm{ref}+} \\ \hline Y_{i|i}^{\sim\mathrm{ref}} \\ \hline Y_{\mathrm{f}}^{-} \end{array}\right) \begin{array}{l} \updownarrow r(i+1) \\ \updownarrow \quad l-r \\ \updownarrow l(i-1) \end{array} \tag{6.119}$$

6.8.2.2 Data-Driven Stochastic Subspace Identification (SSI-DATA)

Recently a lot of research effort in the system identification community has been spent on subspace identification, as evidenced by the book of Van Overschee and De Moor [15] and the second edition of Ljung's book [18]. Subspace methods identify state-space models from (input and) output data by applying robust numerical techniques such as QR factorization, single value decomposition (SVD) and least squares. A general overview of data-driven subspace identification (both deterministic and stochastic) is provided in the book of Van Overschee and De Moor [15]. Although somewhat more involved as compared to previous methods, it is also possible with SSI-DATA to reduce the dimensions of the matrices by introducing the idea of the reference sensors. This is demonstrated in references [19] and [20] and also in this subsection.

The derivation of SSI-DATA is given for the reference sensor case. The original algorithm is simply recovered by considering all sensors as references. First, the Kalman filter states will be introduced because of their importance in subspace identification. Next, the principles of SSI-DATA are explained. Finally, the implementation of the projection in terms of the QR factorization is discussed.

The SSI-DATA method identifies a stochastic state-space model, equation (6.105), from output-only data.

Kalman Filter States

The Kalman filter plays an important role in SSI-DATA. In subsection 6.8.1.3, it was indicated how the forward innovation model equation (6.112) can be obtained by applying the steady-state Kalman filter to the stochastic state-space model equation (6.105). In this section, the non-steady-state Kalman filter is introduced. The Kalman filter is described in many books. A nice derivation can be found in Appendix B of reference [21]. The aim of the Kalman filter is to produce an optimal prediction for the state vector x_k by making use of observations of the outputs up to time $k-1$ and the available system matrices together with known noise covariances. These optimal predictions are denoted by $\hat{x}_{k+1}$. When the initial state estimate $\hat{x}_0 = 0$, the initial covariance of the state estimate $P_0 = E[\hat{x}_0 \hat{x}_0^{\mathrm{T}}] = 0$ and the output measurements $y_0, \ldots, y_k - 1$ are given, the non-steady-state Kalman filter state estimates $\hat{x}_k$ are obtained by the following recursive formulas:

$$\begin{aligned} \hat{x}_k &= A\hat{x}_{k-1} + K_{k-1}(y_{k-1} - C\hat{x}_{k-1}) \\ K_{k-1} &= (G - AP_{k-1}CT)(R_0 - CP_{k-1}C^{\mathrm{T}})^{-1} \\ P_k &= AP_{k-1}A^{\mathrm{T}} + (G - AP_{k-1}C^{\mathrm{T}})(R_0 - CP_{k-1}C^{\mathrm{T}})^{-1}(G - AP_{k-1}C^{\mathrm{T}})^{\mathrm{T}} \end{aligned} \tag{6.120}$$

expressing the Kalman state estimate, the Kalman filter gain matrix and the Kalman state covariance matrix. The Kalman filter state sequence $X_i \in \mathbb{R}^{n \times N}$ is defined as

$$\hat{X}_i = (\hat{x}_i \, \hat{x}_{i+1} \cdots \hat{x}_{i+N-1}) \tag{6.121}$$

The correct interpretation of the $(q+1)$th column of this matrix is that this state $\hat{x}_{i+q}$ is estimated according to equation (6.120) by using only i previous outputs: $y_q, \ldots, y_{i+q-1}$. By consequence, two consecutive elements of $\hat{\mathrm{X}}_i$ cannot be considered as consecutive iterations of equations (6.120). More details can be found in reference [15]. It is important to note that a closed-form expression exists for this Kalman filter state sequence and that it is this sequence that will be recovered by the SSI-DATA algorithm.

Data-driven Stochastic Subspace Identification Theory

The SSI-DATA algorithm starts by projecting the row space of the future outputs into the row space of the past reference sensors. The idea behind this projection is that it retains all the information in the past that is useful to predict the future. The notation and definition of this projection is

$$\mathcal{P}_i^{\mathrm{ref}} = Y_{\mathrm{f}} / Y_{\mathrm{p}}^{\mathrm{ref}} = Y_{\mathrm{f}}(Y_{\mathrm{p}}^{\mathrm{ref}})^{\mathrm{T}}(Y_{\mathrm{p}}^{\mathrm{ref}}(Y_{\mathrm{p}}^{\mathrm{ref}})^{\mathrm{T}})^{\dagger} Y_{\mathrm{p}}^{\mathrm{ref}} \tag{6.122}$$

The matrices $Y_{\mathrm{f}} \in \mathbb{R}^{li \times N}$ and $Y_{\mathrm{p}}^{\mathrm{ref}} \in \mathbb{R}^{ri \times N}$ are partitions of the data Hankel matrix $H^{\mathrm{ref}} \in \mathbb{R}^{(r+l)i \times N}$, as indicated in equation (6.118). Note that expression (6.122) is only the definition of $\mathcal{P}_i^{\mathrm{ref}}$; it does not indicate how the projection is computed. It will be seen later that it is computed by the numerically robust QR factorization.

The main theorem of stochastic subspace identification [15] states that the projection $\mathcal{P}_i^{\text{ref}}$ can be factorized as the product of the extended observability matrix $\boldsymbol{O}_i$ and the Kalman filter state sequence $\hat{\mathbf{X}}_i$ (equation (6.121)):

$$\mathcal{P}_i^{\text{ref}} = \boldsymbol{O}_i\hat{\boldsymbol{X}}_i = \underset{\overset{\leftrightarrow}{n}}{\begin{pmatrix} \boldsymbol{C} \\ \boldsymbol{CA} \\ \cdots \\ \boldsymbol{CA}^{i-1} \end{pmatrix}} (\hat{\boldsymbol{x}}_i\ \hat{\boldsymbol{x}}_{i+1} \cdots \hat{\boldsymbol{x}}_{i+N-1}) \tag{6.123}$$

The proof of this theorem for the case where all outputs are considered as references ($\boldsymbol{Y}_\text{p}^{\text{ref}} \to \boldsymbol{Y}_\text{p}$) can be found in reference [15]. In the present case, where only the past reference outputs have been used, the proof is almost the same, except for the significance of the obtained Kalman filter state sequence $\hat{\mathbf{X}}_i$. The non-steady-state Kalman filter is applied to a reduced state-space model that includes only the reference outputs. The following substitutions have to be made in equation (6.120):

$$\begin{aligned} \boldsymbol{y}_k &\to \boldsymbol{y}_k^{\text{ref}} = \boldsymbol{L}\boldsymbol{y}_k \\ \boldsymbol{G} &\to \boldsymbol{G}\boldsymbol{L}^{\text{T}} \\ \boldsymbol{C} &\to \boldsymbol{L}\boldsymbol{C} \\ \boldsymbol{R}_0 &\to \boldsymbol{L}\boldsymbol{R}_0\boldsymbol{L}^{\text{T}} \end{aligned} \tag{6.124}$$

At first sight, the choice of the reference sensors seems to be unimportant: for all choices factorization (equation (6.123)) is found. Indeed, theoretically the internal state of a system does not depend on the choice and number of observed outputs. However, in identification problems where the states are estimated based on observations, the choice and number of outputs does matter: different reference outputs will lead to different Kalman filter state estimates $\hat{\mathbf{X}}_i$.

Since the projection matrix is the product of a matrix with n columns and a matrix with n rows (equation (6.123)), its rank equals n (if $li \geq n$). The SVD is a numerically reliable tool to estimate the rank of a matrix. After omitting the zero singular values and corresponding singular vectors, the application of the SVD to the projection matrix yields

$$\mathcal{P}_i^{\text{ref}} = \boldsymbol{U}_1\boldsymbol{S}_1\boldsymbol{V}_1^{\text{T}} \tag{6.125}$$

where $\boldsymbol{U}_1 \in \mathbb{R}^{li\times n}$, $\boldsymbol{S}_1 \in (\mathbb{R}_0^+)^{n\times n}$ and $\boldsymbol{V}_1 \in \mathbb{R}^{N\times n}$. The extended observability matrix and the Kalman filter state sequence are obtained by splitting the SVD into two parts:

$$\begin{aligned} \boldsymbol{O}_i &= \boldsymbol{U}_1\boldsymbol{S}_1^{1/2}\boldsymbol{T} \\ \hat{\boldsymbol{X}}_i &= \boldsymbol{O}_i^{\dagger}\mathcal{P}_i^{\text{ref}} \end{aligned} \tag{6.126}$$

Up to now the order of the system n (as the number of non-zero singular values in equation (6.125)), the observability matrix $\boldsymbol{O}_i$ and the state sequence $\hat{\mathbf{X}}_i$ were found. However, the identification problem is to recover the system matrices $\boldsymbol{A}$, $\boldsymbol{G}$, $\boldsymbol{C}$ and $\boldsymbol{R}_0$. If the separation between past reference and future outputs in the Hankel matrix is shifted one block row down, as indicated in equation (6.119), another projection can be defined as

$$\mathcal{P}_{i-1}^{\mathrm{ref}} = \boldsymbol{Y}_{\mathrm{f}}^{-} / \boldsymbol{Y}_{\mathrm{p}}^{\mathrm{ref}+} = \boldsymbol{O}_{i-1}\hat{\boldsymbol{X}}_{i+1} \tag{6.127}$$

where the proof of the second equality is similar to the proof of the main subspace theorem (6.123). The extended observability matrix $\boldsymbol{O}_{i-1}$ is simply obtained after deleting the last l rows of $\boldsymbol{O}_i$:

$$\boldsymbol{O}_{i-1} = \boldsymbol{O}_i(1 : l(i-1), :) \tag{6.128}$$

The state sequence $\hat{\mathbf{X}}_{i+1}$ can now be computed as

$$\hat{\boldsymbol{X}}_{i+1} = \boldsymbol{O}_{i-1}^{\dagger}\mathcal{P}_{i-1}^{\mathrm{ref}} \tag{6.129}$$

At this moment the Kalman state sequences $\hat{\mathbf{X}}_i$ and $\hat{\mathbf{X}}_{i+1}$ are calculated using only the output data. The system matrices can now be recovered from the following overdetermined set of linear equations, obtained by stacking the state-space models for time instants i to $i+N_{-1}$:

$$\begin{pmatrix} \hat{\boldsymbol{X}}_{i+1} \\ \boldsymbol{Y}_{i|i} \end{pmatrix} = \begin{pmatrix} \boldsymbol{A} \\ \boldsymbol{C} \end{pmatrix} \hat{\boldsymbol{X}}_i + \begin{pmatrix} \boldsymbol{W}_i \\ \boldsymbol{V}_i \end{pmatrix} \tag{6.130}$$

where $\boldsymbol{Y}_{i|i} \in \mathbb{R}^{l \times N}$ is a Hankel matrix with only one block row (6.118) and $\boldsymbol{W}_i \in \mathbb{R}^{n \times N}$, $\boldsymbol{V}_i \in \mathbb{R}^{l \times N}$ are the residuals. Since the Kalman state sequences and the outputs are known and the residuals are uncorrelated with $\hat{\mathbf{X}}_i$, the set of equations can be solved for $\boldsymbol{A}$ and $\boldsymbol{C}$ in a least squares sense:

$$\begin{pmatrix} \boldsymbol{A} \\ \boldsymbol{C} \end{pmatrix} = \begin{pmatrix} \hat{\boldsymbol{X}}_{i+1} \\ \boldsymbol{Y}_{i|i} \end{pmatrix} = \hat{\boldsymbol{X}}_i^{\dagger} \tag{6.131}$$

The noise covariances $\boldsymbol{Q}$, $\boldsymbol{R}$ and $\boldsymbol{S}$ are recovered as the covariances of the least squares residuals:

$$\begin{pmatrix} \boldsymbol{Q} & \boldsymbol{S} \\ \boldsymbol{S}^{\mathrm{T}} & \boldsymbol{R} \end{pmatrix} = \begin{pmatrix} \boldsymbol{W}_i \\ \boldsymbol{V}_i \end{pmatrix} = (\boldsymbol{W}_i^{\mathrm{T}} \boldsymbol{V}_i^{\mathrm{T}}) \tag{6.132}$$

From the properties of stochastic systems, it is easy to see how the matrices $\boldsymbol{A}$, $\boldsymbol{C}$, $\boldsymbol{Q}$, $\boldsymbol{R}$ and $\boldsymbol{S}$ can be transformed to $\boldsymbol{A}$, $\boldsymbol{G}$, $\boldsymbol{C}$ and $\boldsymbol{R}_0$. First the Lyapunov equation is solved for Σ:

$$\Sigma = \boldsymbol{A}\Sigma\boldsymbol{A}^{\mathrm{T}} + \boldsymbol{Q} \tag{6.133}$$

after which $\boldsymbol{G}$ and $\boldsymbol{R}_0$ can be computed as

$$\begin{aligned} \boldsymbol{R}_0 &= \boldsymbol{C}\Sigma\boldsymbol{C}^{\mathrm{T}} + \boldsymbol{R} \\ \boldsymbol{G} &= \boldsymbol{A}\Sigma\boldsymbol{C}^{\mathrm{T}} + \boldsymbol{S} \end{aligned} \tag{6.134}$$

At this point the identification problem is theoretically solved: based on the outputs, the system order n and the system matrices $\boldsymbol{A}$, $\boldsymbol{G}$, $\boldsymbol{C}$ and $\boldsymbol{R}_0$ are found.

The matrices $\boldsymbol{A}$ and $\boldsymbol{C}$ are sufficient to compute the modal parameters. The discrete poles Λ_{d} and the observed mode shapes $\boldsymbol{V}$ are computed as

$$\begin{aligned} \boldsymbol{A} &= \Psi\Lambda_{\mathrm{d}}\Psi^{-1} \\ \boldsymbol{V} &= \boldsymbol{C}\Psi \end{aligned} \tag{6.135}$$

REFERENCES

1. Cooley, J.M. and Tukey, J.W. (1965) An algorithm of the machine calculation of complex Fourier series. *Mathematics of Computation*, **19**(90), 297–301.
2. Clough, R.W. and Penzien, J. (1993) *Dynamics of Structures*, 2nd edn, McGraw-Hill, New York.
3. Adam, Ch. (1996) *Rechenübungen aus Baudynamik*. Institut für Allgemeine Mechanik der Technischen Universität, Wien.
4. Cantieni, R. *et al.* (1996) *Untersuchung des Schwingungsverhaltens großer Bauwerke*, Technische Akademie, Esslingen.
5. Eibl, J. *et al.* (1988) Baudynamik, in *Betonkalender 1988*, Band II, Ernst & Sohn, Berlin.
6. Agrati, S. (1994) *Estimation of Structural Parameters from Ambient Vibration Test*. Master thesis, Danish Technical University.
7. Mehrabi, A.B. and Tabatabai, H. (1998) A unified finite difference formulation for free vibration of cable, *Journal of Structural Engineering*, **124**(11) 1313–22.
8. Forstner, E. (2003) *Experimentelle Untersuchung des Schwingverhaltens externer Spannglieder*. Diploma thesis, Institut für Konstruktiven Ingenieurbau, Vienna.
9. De Roeck, G. and Van Gysel, E. (2002) *Estimation of cable tension from vibrations*. IMAC Project, January.
10. De Roeck, G. and Van Gysel, E. (2002) *Estimation of cable forces and bending stiffness*. IMAC Project, April.
11. Stöcker, H. (1999) *Taschenbuch mathematischer Formeln und moderner Verfahren*, Verlag Harri Deutsch, Thun/Frankfurt, Germany.
12. Savov, K. and Wenzel, H. (2004) *System Identification of a Steel Frame: Comparison between Frequency and Time Domain Analysis*. Submitted to 11th Workshop of the

European Group for Intelligent Computing in Engineering, 31 May–1 June 2004, Weimar, Germany.
13. Chopra, A.H. (2001) *Dynamics of Structures: Theory and Applications to Earthquake Engineering*, 2nd edn, Prentice-Hall, Englewood Cliffs, New Jersey.
14. Peeters, B. (2000) *System Identification and Damage Detection in Civil Engineering*. PhD thesis, Katholieke Universiteit Leuven, Leuven, Belgium, December.
15. Van Overschee, P. and De Moor, B. (1996) *Subspace Identification for Linear Systems: Theory – Implementation – Applications*. Kluwer Academic Publishers, Dordrecht, The Netherlands.
16. Ewins, D.J. (1984) *Modal Testing: Theory and Practice*. Research Studies Press Ltd, Letchworth, Hertfordshire.
17. Heylen, W., Lammens, S. and Sas, P. (1995) *Modal Analysis Theory and Testing*. Department of Mechanical Engineering, Katholieke Universiteit Leuven, Leuven, Belgium, 1995.
18. Ljung, L. (1999) *System Identification: Theory for the User*, 2nd edn, Prentice-Hall, Upper Saddle River, New Jersey.
19. Peeters, B. and De Roeck, G. (1999) Reference-based stochastic subspace identification for output-only modal analysis. *Mechanical Systems and Signal Processing*, **13**(6), 855–878.
20. Peeters, B. and De Roeck, G. (1999) *Reference Based Stochastic Subspace Identification in Civil Engineering*. Proceedings of the 2nd International Conference on Identification in Engineering Systems, March 1999, Swansea, pp. 639–48.
21. Juang, J.-N. (1994) *Applied System Identification*, Prentice-Hall, Englewood Cliffs, New Jersey.

7

Outlook

The development potential of the method is huge, in particular with regard to the increasingly improving possibilities of measuring technology and computer capacity. Accordingly, such systems are developed worldwide. For final acceptance of the method it will be necessary to intensify the training in issues of construction dynamics and transport information on the method. Acceptance will be gained if the application of knowledge-based systems and probabilistic considerations are given more attention in structural assessment. By means of approaches from soft computing methods it will be possible to identify damage automatically. This will, however, by no means replace the judgement of the bridge engineer in the foreseeable future. Therefore it seems to be extremely important that the method is operated by bridge engineers who acquire knowledge on measuring technology, particularly in the years of its development. For the engineer this is a tool for the quantification of several subjectively noticeable phenomena.

The desired targets for the development in the near future are:

- the development of a system without data cables and new sensors;
- the development of efficient algorithms for the evaluation of measurements;
- the formulation of damage scenarios and algorithms for the automatic recognition of the latter;
- the integration of the method in decision support systems (knowledge-based);
- pattern recognition software in conjunction with sensor networks.

Despite the high potential of the method it must be pointed out that it cannot replace but only complement traditional methods. A further development of traditional methods is enabled. A far-reaching combination between visual inspection, conventional inspection methods and dynamic vibration analysis is advisable.

Ambient Vibration Monitoring H. Wenzel and D. Pichler

7.1 DECISION SUPPORT SYSTEMS

Decision support systems will have to be furnished with levels of decisions to be taken. The condition rating might be done with a traffic light indication for easy recognition.

1. *Normal operation.* A cost-effective monitoring system should be implemented. A normal sampling and transmission rate is chosen. Operator information is provided on a daily basis. The condition rating is green.
2. *Low margin operation.* In general, the same applies as for level 1, but the sampling rate should be increased as well as the information to the operator. The condition rating is still green.
3. *Warning level.* The monitoring system should be upgraded to the necessary level. The sampling rate is increased and frequent information is transferred to the operator. The condition rating shows yellow.
4. *Emergency level.* The monitoring system should be directly connected to user interfaces such as traffic lights. An automatic shutdown is considered in the case of emergency. The monitoring system is extended to the maximum possible size and permanent monitoring is conducted. A display of the results is done on a permanent level. The condition rating is red.
5. *Post mortem level.* Data of the incident should serve an effort to understand how the phenomena happened. Like a black box of airplanes all data at the time of occurrence are recorded and stored for evaluation. A procedure to make the best possible use of the data is defined.

The levels could be demonstrated in comparison to a typical stress–strain curve that is valid for steel. Normal operation represents the area up to 50% of ultimate strength; a low margin operation means any operation close to the limits of the standards.

The warning level could be triggered when the highest value for exceptional load cases have been reached (i.e. 0.8 of ultimate strength). The emergency level could represent the yielding point of the curve minus 5% (or any other fractile). Each of the levels requires different instrumentation, different operation and different actions by the owner. Strategies for each of the levels should be developed.

7.2 SENSOR TECHNOLOGY AND SENSOR NETWORKS

7.2.1 STATE-OF-THE-ART SENSOR TECHNOLOGY

Sensors and associated technologies have significantly evolved over the last three decades, motivated by seeking new solution approaches to optimal performance and safety protection of civil infrastructures. While the physics behind the transduction process may have not achieved major and revolutionary advances, the overall sensing systems technology has experienced major strives for further enhancement. Over the time, sensor technology has progressed on three levels:

- its focus;
- its level of intelligence;
- its architecture.

Most of today's application areas, especially in infrastructure systems, are using the sensor technology that is characterized by focused application, local intelligence and hierarchical architecture. However, it is the goal of new research and development, projects and plans to move the technology towards the next generation of sensor systems, which will have the following characteristic:

- ubiquitous sensing, incorporating low costs, low power, easily deployable, wireless and miniaturized in size as their unique properties;
- agent-based sensing configuring distributed intelligence through integration of agents that would support sensor correlation and collaboration, emergent behaviour and graceful degradation;
- anticipatory prognostics intelligence integration within data processing to anticipate potential faults in advance of experiencing failure patterns.

In order to realize ubiquitous sensing, several enabling technologies have to be developed and reach their mature stage. These include:

- MEMS- and NEMS (micro/nanoelectromechanical systems)-based low-cost transducers;
- goal-focused, self-aware sensor intelligence to convert data to useful information that will provide efficient knowledge;
- high-performance low-cost computation technology;
- in expensive wireless communication technology for harsh environment;
- minimum power dependence and energy sources capable of scavenging power from the environment.

Considering the required characteristics associated with the next generation of the sensor technology. It is clear that new research and development has to be rigorously initiated in order to meet the challenges in the future.

7.3 RESEARCH GAPS AND OPPORTUNITIES

Sensors have been used for detection and measurement of various kinds of data and quantities such as temperature, strain, pressure and other dynamic force and response environmental parameters. They are developed through multiple disciplinary sources such as materials and engineering in all fields. Sensors are also found being used in many diverse situations, such as to monitor and diagnose, provide surveillance and status, evaluate and improve performance, derive control actions, protect from damage and failure, get information and enhance intelligence, etc.

Through recent R&D programs driven by societal demands and motivated by technological opportunities new sensor techniques have been developed across a wide spectrum including, for example, acoustic sensing, radiometry, electronic and nuclear imaging, chemical species selective sensing, optic sensors, MEMS and NEMS sensing, etc. From these techniques it can be recognized that sensors and sensor networks do not represent a standalone technology but rather are at the interface of some of the dominant transcendence technologies of the current work, namely:

- wireless technology, use of radio telemetry, cellular dialup, satellite, etc., for broad deployment of sensors and data transmission;
- information technology processing and diagnosis of large volumes of continuous on-line data communication between self-reconfiguration of sensors, uniformity in condition identification and damage/failure/emergency countermeasure flagging, etc.
- communication and networking technology such as enabling regional and global data recording, flow, and distribution and monitoring.

Much progress has been made in many fronts of sensors in the current rapid transition to a data-rich world from the past data-poor world, many research gaps still exist that call for attention. These are namely:

New sensor design including:

- improved resolution, sensitivity;
- hybrid sensing (bio- and chemical sensing, for example);
- increased bandwidths (terahertz);
- new sensing, actuating materials;
- nano-imaging;
- packaging and fabricating.

Sensor informatics:

- intelligent, reconfigurable sensors;
- data management and reconfiguration algorithms;
- signal-to-noise enhancement;
- self-recalibration;
- networking of densely embedded sensors.

Power and data (in/out):

- size;
- embedded use;
- talking among and movement of nano-sensors.

A sensor network of structures would be comprised of a large number of sensor nodes, which are densely deployed on or inside the structure, and strategically placed or embedded. There are several factors and challenges that influence successful deployment and operation of sensor networks within a structure. These include:

- fault tolerance;
- scalability;
- accessibility;
- manufacturing costs;
- operating environment;
- sensor network topology;
- hardware constraints;
- transmission media;
- power consumption.

Many researchers have addressed some of these factors. However, none of these studies has a full integrated view of all factors that are driving the design of sensor networks and sensor nodes. Therefore new research is needed to address these challenges, which have a profound impact on feasibility and deployment of structural sensor networks and thus realization of smart structures.

7.4 INTERNATIONAL COLLABORATION

The objectives intend to contribute to enhance international collaboration between key institutions, academia and industry to share knowledge, methodologies, tools and results of past and future research. International collaboration shall be installed through networking, joint participation to international projects, exchange of researchers and contribution to national projects of common interest.

It is recognisable that the will for international collaboration exists in the engineering community. Nevertheless the basis for successful work lax the necessary tools and an initiative which is determined to actively pursue the objectives.

7.4.1 COLLABORATION FRAMEWORK

The following framework is based on meetings and discussions performed by the international community during events held in 2003. This framework will be dynamically developed.

7.4.1.1 Topics of Common Interest

The following important issues will be addressed for a successful collaboration at the international level:

- international agreement;
- codes and standards;
- common database and data format, metadata protocol;
- data interrogation;
- networking;
- system integration;
- education and training;
- exchange of researchers;
- management of the collaboration;
- funding.

7.4.1.2 Issues of Structural Health Monitoring, Structural Control and Performance-Based Design of Structural and Non-Structural Systems

The following general needs will be covered by the initiative:

- novel sensors and sensor networks;
- appropriate level of instrumentation;
- communication and wireless transmission;
- data mining;
- performance diagnosis;
- interdisciplinary and collaborative research.

In addition, the following specific issues should be addressed:

- development of benchmark models for SHM, with emphasis on nonlinear system identification and damage diagnosis, and verification of computation ability;
- international collaboration on the establishment of testing procedures to generate datasets such as data on the dynamic response of bridges and buildings;
- design of data acquisition systems for large-scale tests, taking into account the need to exploit the data for structural health monitoring research;
- harmonization of codes, guidelines and data format.

Regarding the research on structural control, the following topics have been considered as highly relevant:

- development of benchmark models for control research, and their use for a cost–benefit analysis of the different control devices and algorithms;
- implementation issues;

- harmonization of design guidelines;
- research on the control of non-structural components;
- integration of structural health monitoring and control;
- development of low-cost sensors and devices.

Regarding the performance-based design of structural and non-structural systems, the following issues were identified as most important:

- performance-based design criteria for integrated structural/non-structural systems;
- methodology for performance assessment;
- ground motion evaluation procedures for performance-based design;
- development of fragility curves for non-structural systems;
- application of structural control to mitigate the earthquake hazard on hospitals as well as on ordinary house designs.

7.4.1.3 Issue Networked Testing and Structuring Retrofit

The following common action should be developed for the following topics to set up an effective collaboration at international level:

- data archiving;
- collaborative tools;
- international networked testing;
- seismic retrofit;
- exchange of researchers;
- versatility of infrastructure;
- resolutions.

Data archiving includes the content, data format and metadata. It should include the following items:

- material data, calibration data and raw data;
- calibrated data;
- photos and videos;
- project description and design drawings;
- numerical simulation data and software version;
- project management data;
- data-sharing protocol;
- NEES (network for earthquake engineering simulation) data format;
- data searching tools.

In the common software framework the following collaborative tools should be included:

- CHEF (CompreHensive collaborativE Framework, developed at the University of Michigan, USA and being used by NEESgrid);
- ISEE (Internet-based Simulation for Earthquake Engineering, National Center for Research on Earthquake Engineering, Taiwan);
- planning tools;
- browsers;
- searching tools;
- message board and chatting tools;
- electronic notebook;
- data display;
- file repository;
- open source.

The international networked testing should involve:

- sharing experience and methodology;
- exploitation of complementary facilities;
- increase in the scale of projects;
- education to researchers and engineers;
- large-scale, complex or new structural system and component tests.

For the topic of seismic retrofit, the specific targets include masonry buildings, non-ductile structures and non-structural components. The issues to be addressed include:

- large-scale or complex system evaluation;
- new material and devices;
- validation of numerical models.

The exchange of researchers should be facilitated through:

- creation of a database for exchange mechanism of researchers and students;
- tele-participation;
- co-supervision of PhD students.

Advantage should be taken of the versatility of the available infrastructures through common use of existing competences and development in different areas:

- real-time pseudo-dynamic test technology and substructuring;
- servo-control technology;
- full-scale testing;
- shake table/actuator hybrid testing;
- complementary facilities and competencies;
- information technology;

- sensor technology;
- computing technology;
- simulation tools.

7.4.2 *ACTIVITIES*

Based on the above-mentioned framework the following activities should be pursued:

- establishment of specific international collaboration agreements among global players in the field;
- building up of a researcher's network;
- organization of a work group on data formatting;
- organization of a work group on data archiving;
- setting up of a database for exchange mechanism of researchers and students;
- establishment of research programmes, including international collaboration;
- any other activity suitable to achieve the objectives.

This plan shall be maintained and further developed.

8

Examples for Application

8.1 AITERTAL BRIDGE, POST-TENSIONAL T-BEAM (1956)

Location:	Linz, Upper Austria, Austria
Client:	Government of Upper Austria
Checking period:	1997, 1998

The Aitertal Bridge (Figure 8.1) has an overall length of 428 m. In 1997 measurements of the dynamic behaviour and the global actual condition of the bridge were carried out together with a visual inspection. The purpose of the tests of 1997 was to assess the condition of the structure in independent ways. In 1998 both structures were measured again and the results were compared with the measurements of 1997 (Figures 8.2 to 8.4).

The objective of the tests carried out on the Aitertal Bridge was to ascertain and document the condition of the bridge and its development. For this reason investigations were conducted over a period of two years, which clearly showed the annual developments.

The assessment of its vibration behaviour is useful to determine the load-bearing capacity and security of the structure. The measurements were carried out on both structures (north and south) as well as on their bearing on columns and bearing pads. The substructure, namely columns and bearing pads, and the foundation were not part of the analysis.

Ambient Vibration Monitoring H. Wenzel and D. Pichler

Figure 8.1 Aitertal Bridge

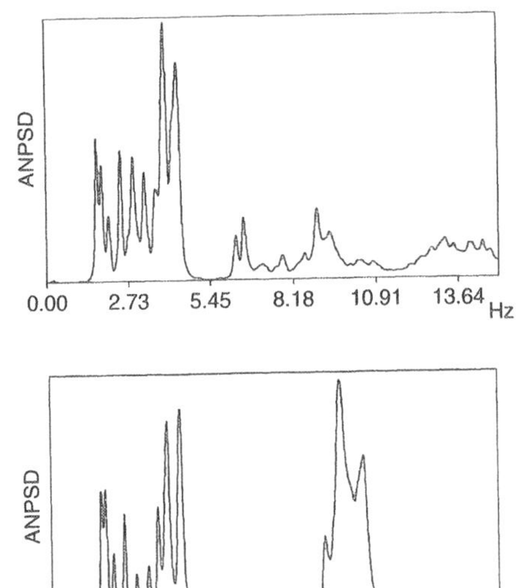

Figure 8.2 ANPSD structure south, 1997 and 1998

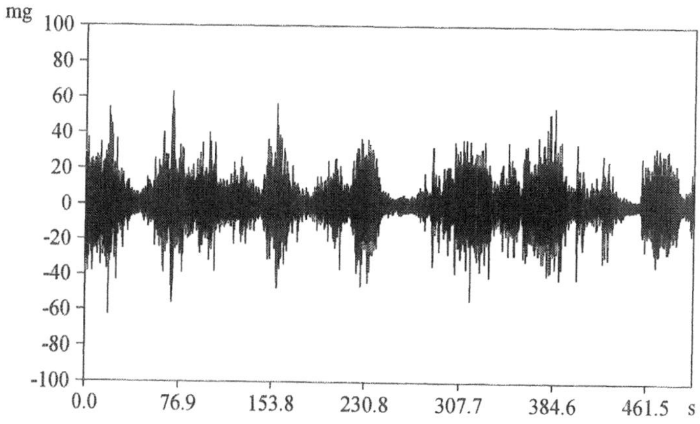

Figure 8.3 Acceleration

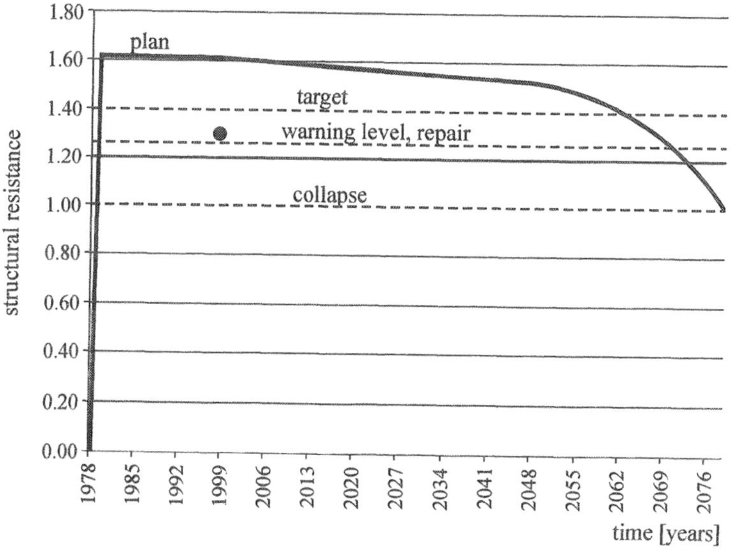

Figure 8.4 Life cycle

8.2 DONAUSTADT BRIDGE, CABLE-STAYED BRIDGE IN STEEL (1996)

Location:	Vienna, Austria
Client:	Municipal authorities, Vienna
Construction period:	1995–1996
Checking period:	1998, 2001

The new Donaustadt Bridge (Figure 8.5) is 340 m long with a main span length of 186 m. It was designed as a cable-stayed bridge with one 75-m-high A-shaped pylon made of steel and with steel girders. The 370-m-long structures over the Danube Island and the New Danube are pre-stressed concrete bridges.

Measurements of the dynamic behaviour are carried out to assess the actual structural systems for the construction stages and for the final system. By doing several ambient vibration tests and transforming the measured data into power spectrums the dynamic parameters of the structural members were identified (Figures 8.6 and 8.7).

Figure 8.5 Donaustadt Bridge

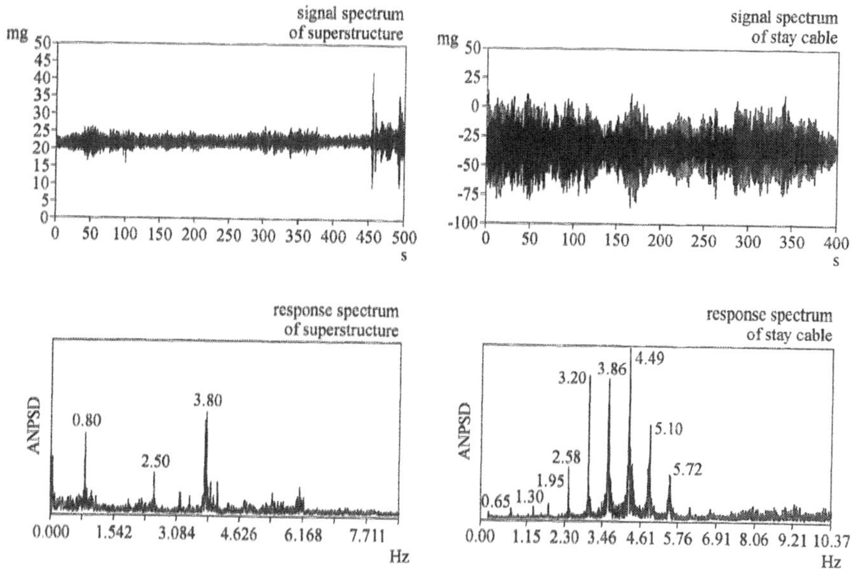

Figure 8.6 Measurements of natural frequencies under regular traffic

Figure 8.7 Sensor mounted on a cable

8.3 F9 VIADUCT DONNERGRABEN, CONTINUOUS BOX GIRDER (1979)

Location: Salzburg, Austria
Client: Government of Salzburg
Checking period: 1999

The Donnergraben Bridge (Figure 8.8) on the A10 Tauern motorway is executed for each lane as a seven-span pre-stressed concrete structure with a total length of 425.15 m. The single span widths are about 5 × 67.45 m and 2 × 57.45 m (for the downhill structure) and 5 × 69.0 m and 2 × 58.00 m (for the uphill structure). The width of one system is 14.5 m with a constant height of the box girder of 4.60 m and was built by cantilever erection. The building was completed in 1979. Measurements of the dynamic behaviour of the two structural systems were carried out as well as an analysis of the actual general condition of the bridge and to set up a basis to take measures for investment and maintenance (Figures 8.9 to 8.11). Another evaluation of the mode of operation of cracks in the end spans and the effective response of the structure in comparison to the proposed response should follow.

Figure 8.8 Donnergraben Bridge

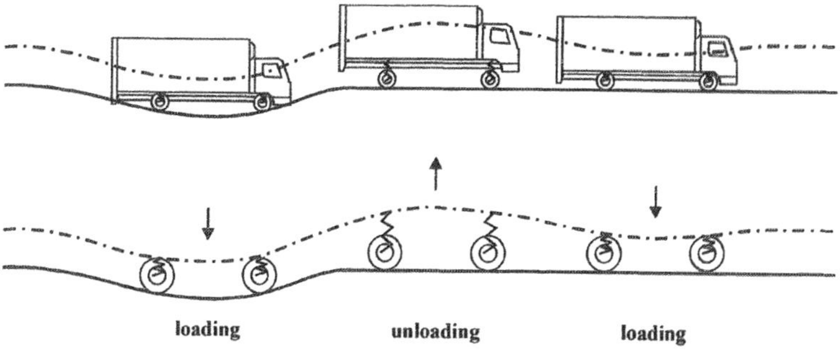

Figure 8.9 High acceleration level caused by pavement damage

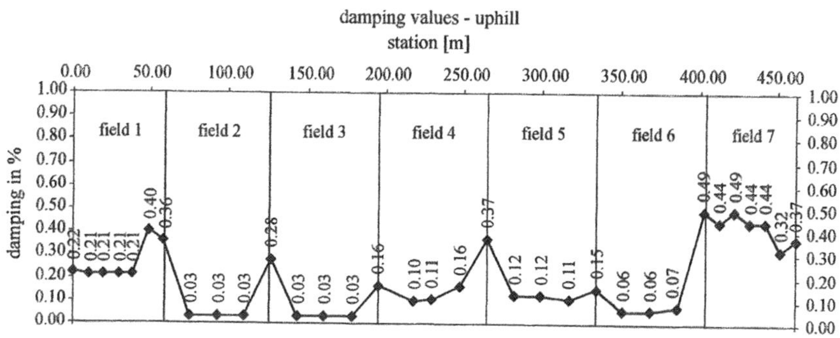

Figure 8.10 Higher damping values caused by cracked spans

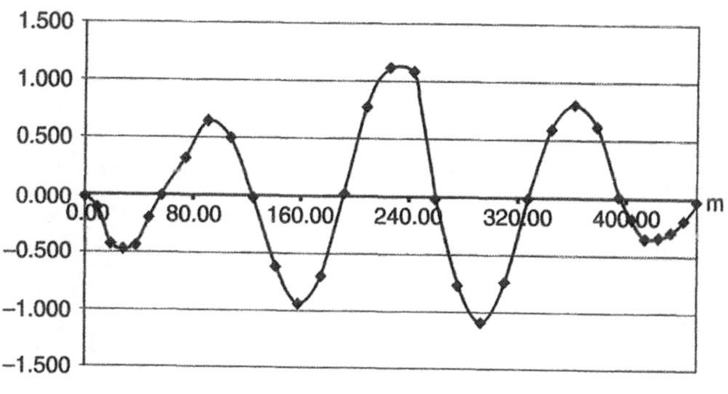

Figure 8.11 First vertical mode shape of the structure

8.4 EUROPA BRIDGE, CONTINUOUS STEEL BOX GIRDER (1961)

Location:	Innsbruck, Tyrol, Austria
Operator:	ASAG–Alpenstrassen AG, Austria
Start of SHM:	May 1998
Structure category:	large-span bridge
Spans:	6 spans: 81/ 108/ 198/ 108/ 81/ 81 m
Structural system:	steel box girder with orthotropic deck and concrete columns
Number of sensors installed:	16
Instrumentation design by:	VCE, (Vienna Consulting Engineers), Austria

The Europa Bridge (Figures 8.12 and 8.13) near Innsbruck, Austria, opened in 1963, is one of the main alpine north–south routes for urban and freight traffic. Currently the bridge is stressed by more than 30 000 motor vehicles per day. The combination of measuring and analytical calculation over the past years has led to detailed system identification. Due to the requirement to assess the prevailing vibration intensities with regard to fatigue problems and possible damage, a permanent measuring system was installed in 2003.

The superstructure is represented by a steel box girder (variable height along the bridge-length) and an orthotropic deck and bottom plate. This motorway bridge with a total length of 657 m comprises six lanes, three for every direction distributed on a width of almost 25 m.

Figure 8.12 Europa Bridge, Innsbruck, Austria

Figure 8.13 Europa Bridge, Innsbruck, Austria

The superior goal is to determine the relation between the randomly induced traffic loads (vehicles per day) and the fatigue-relevant, dynamic response of the structure. As lifetime predictions in modern standards depend on many assumptions, the emphasis is to replace all these guesstimates by measurements. In that context the focus will be on three ranges:

- global behaviour in dependence of all relevant loading cases;
- cross-sectional behaviour under special consideration of the cantilever regions;
- local systems analysing the interaction between tyres and beam–slab connections.

In each of these levels of analysis the consumption of the structure's overall capacity per year is to be determined. The bridge is permanently equipped with the measurement instruments listed in Tables 8.1, 8.2 and 8.3.

Table 8.1 Sensor details

Type of sensors	Number	Location
Displacement sensors	2	At both abutments
One-dimensional acceleration transducer	3	At the cross-section's cantilevers
Three-dimensional acceleration transducer	3	Orthotropic bottom plate
Wind sensors	1	5 m above the road surface
Temperature sensors	7	Inside and outside the box girder

Table 8.2 Measurement equipment and data management

Type of system	Data management
PC- and standalone-based measuring system	Storage in a long-term database on site Analysis (statistics, frequency analysis, etc.) and graphical presentation and documentation in office Notification via a modem about the successful operation of the measuring system

Table 8.3 Data analysis procedures

Type of analysis	Software	Additional features
Statistics, ambient analysis, rainflow analysis, damage detection and lifetime calculations	Self-made software	No expert system

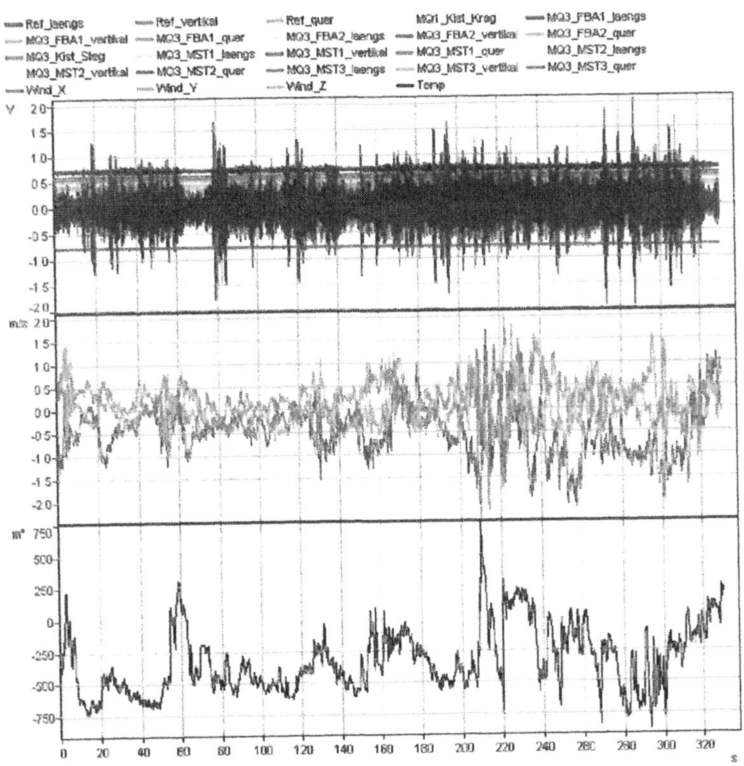

Figure 8.14 Overview of the output of VCE's measuring system

Examples of outcomes:

An indispensable requirement is to reduce the data of the complete load–time history to a few statistical data (rainflow counting) describing the remaining fatigue-relevant recurring response cycles in different categories of intensity and occurrence. Subsequently, all these categories will be subjected to global and local finite element models for lifetime calculations based on the damage-accumulation technique of Palgrem–Miner in connection with *stress-life Wöhler curves* (S–N curves) (see section 4.6.3).

Benefits of using permanent measuring system in the project:

The ability to merge high-precision sensor data of accelerations and displacements in dependence of separately registered wind and temperature data provides the possibility to realize lifetime considerations, which are also of eminent importance for bridge operators (Figures 8.14 and 8.15).

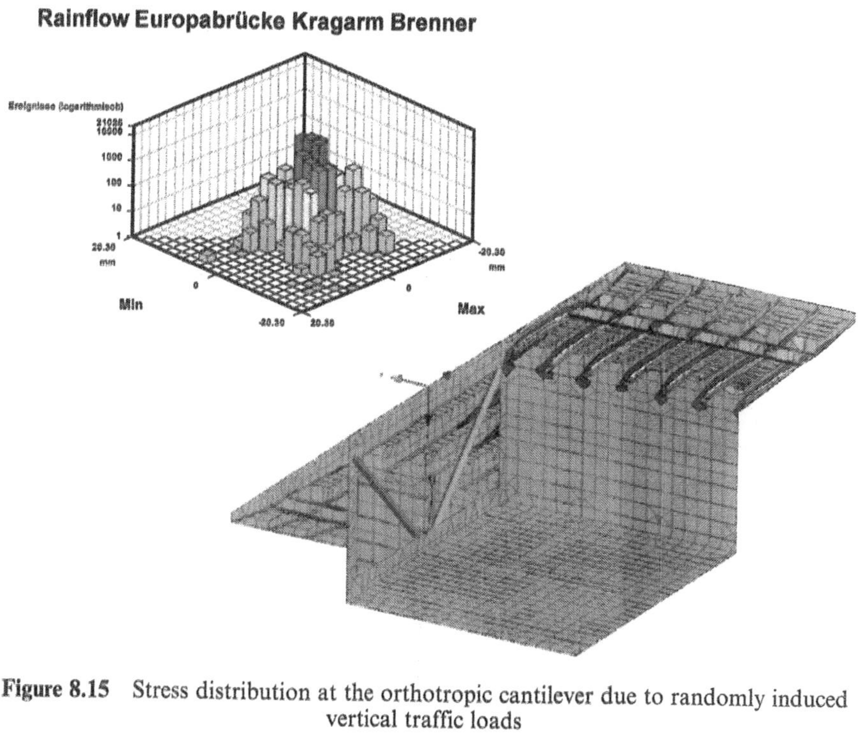

Figure 8.15 Stress distribution at the orthotropic cantilever due to randomly induced vertical traffic loads

8.5 GASTHOFALM BRIDGE, COMPOSITE BRIDGE (1979)

Location:	Salzburg, Austria
Client:	Tauernautobahn AG
Checking period:	1997–1999

The Gasthofalm Bridge (Figure 8.16) is a reinforced concrete bridge with a total length of 436.0 m. It is based on seven spans, the open cross-sections of which are 5 × 66.06 m and 2 × 52.82 m.

Investigations were carried out over several years in order to analyse the condition of the bridge, its development and maintenance over time. Because of uncertainties concerning the effect of the pavement on the system, the bridge was measured in 1998, again after removing the pavement, and the results were compared with those of the basic measurements of 1997 (with pavement). Final measurements were carried out in 1999 (with new pavement) and the effects on the modal parameters were evaluated (Figure 8.17 and Table 8.4).

Figure 8.16 Gasthofalm Bridge

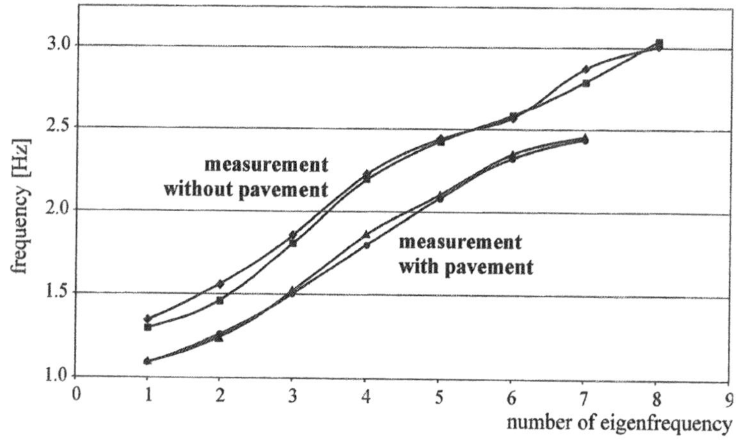

Figure 8.17 Comparison between measured and calculated results of the structure

Table 8.4 Number of eigenfrequency

	Without covering		With covering	
Number	Measurement (Hz)	Calculation (Hz)	Measurement (Hz)	Calculation (Hz)
1	1.30	1.35	1.11	1.10
2	1.46	1.55	1.26	1.26
3	1.81	1.86	1.53	1.51
4	2.19	2.22	1.87	1.80
5	2.42	2.44	2.08	2.08
6	2.58	2.57	2.43	2.33
7	2.79	2.87	2.59	2.44
8	3.04	3.01	2.90	—

8.6 KAO PING HSI BRIDGE, CABLE-STAYED BRIDGE (2000)

Location:	Taiwan, ROC
Client:	TANEEB
Construction period:	1992–1999
Checking period:	2000

Measurements of the dynamic behaviour are carried out to determine the actual load-bearing behaviour of the Kao Ping Hsi Bridge (Figure 8.18) during the construction phases and in the final state (Figures 8.19 and 8.20). The measuring system (Figure 8.21) is able to record all changes of the cable forces during the various construction phases. Influences like the redistribution of the cable forces by effects like creep and shrinking can be assessed.

Figure 8.18 Kao Ping Hsi Bridge

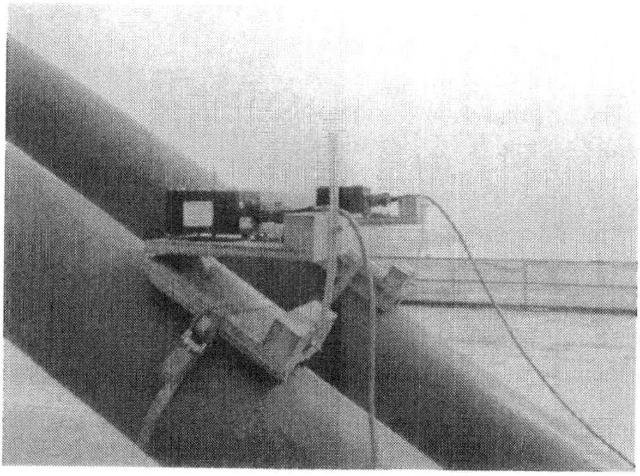

Figure 8.19 Three-dimensional accelerometer on the stay cables

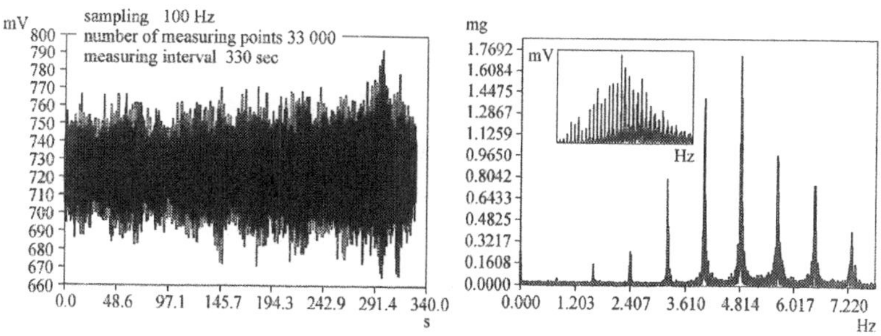

Figure 8.20 Vertical measuring signal (330 s) and vertical frequency spectrum (0–8 Hz)

Figure 8.21 BRIMOS® measuring facility in operation

8.7 INN BRIDGE ROPPEN, CONCRETE BRIDGE (1936)

Location: Tyrol, Austria
Client: Government of Tyrol
Checking period: 1998

The Inn Bridge Roppen (Figure 8.22) represents an arched bridge with an overall length of 165.5 m. The principal component is the 72.5 m long concrete arch. The single spans of the remaining structure amount to 2 × 18 m and 3 × 19 m.

Measurements of the dynamic behaviour of the structural system of the Inn Bridge Roppen were carried out as well as an analysis of the actual general condition of the bridge. The arched structure and both access structures (Telfs and Imst) and their bearing on columns and bearing pads were tested, as illustrated in Figures 8.23 to 8.25. The arch was tested separately. The substructures, i.e. columns and bearing pads, and their foundations were not part of the analysis.

Figure 8.22 Inn Bridge Roppen

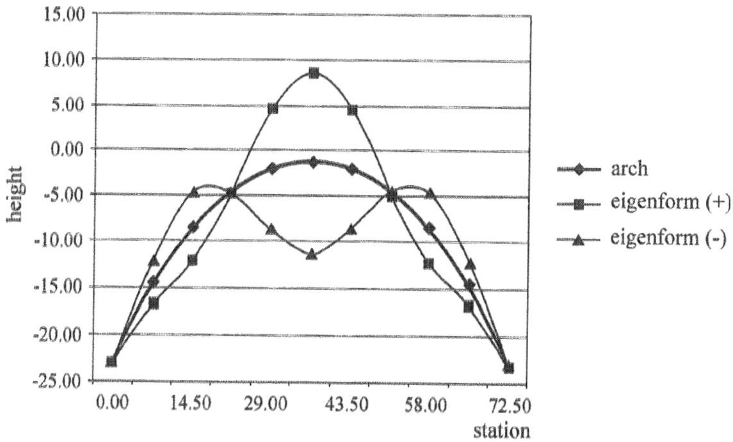

Figure 8.23 Mode shape of the arch

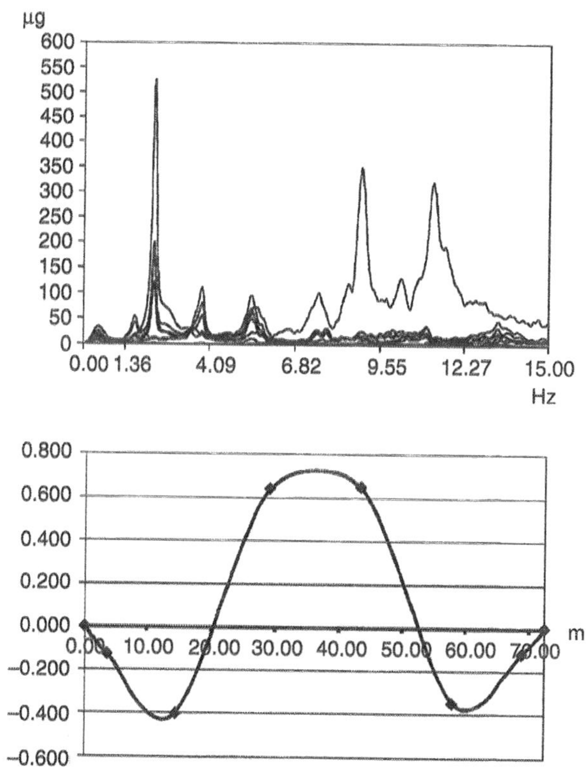

Figure 8.24 Smoothed spectrum of arch and first vertical mode shape of the main structure (right)

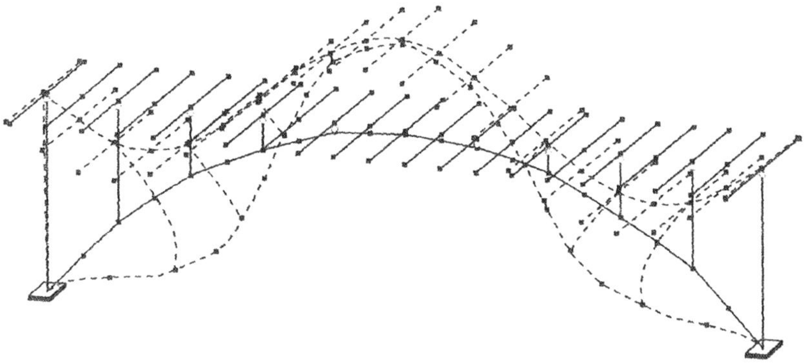

Figure 8.25 First vertical mode shape of the main structure

8.8 SLOPE BRIDGE SAAG, BRIDGE REHABILITATION (1998)

Location:	Carinthia, Austria
Client:	Government of Carinthia
Checking period:	1998–1999

The measurements of the system of the slope bridge Saag (Figure 8.26) were carried out in order to estimate the influence of construction works on the bridge. Owing to very high dynamic stress resulting from the redevelopment of the uphill structure (maximum value 0.81 g), a permanent supervision of the construction works that were carried out on the downhill structure was planned. The aim was to avoid an overload of the structure system as a result of demolition work (Figures 8.27 and 8.28). For this reason an acceleration limit (0.45 g) was introduced and its observance in the course of the construction works was checked.

Figure 8.26 Slope bridge Saag

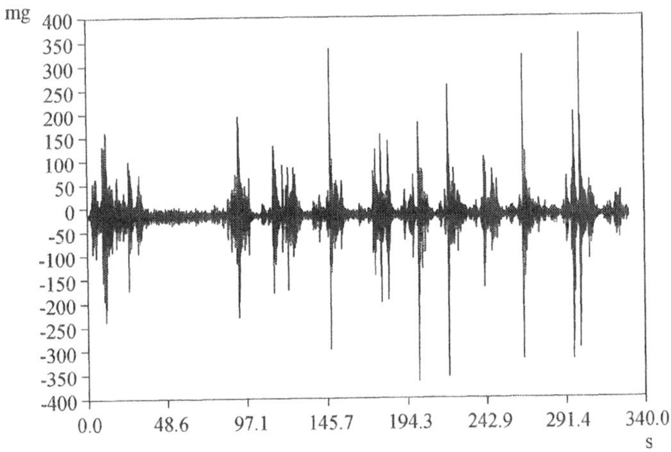

Figure 8.27 Removement of the pavement with an excavator

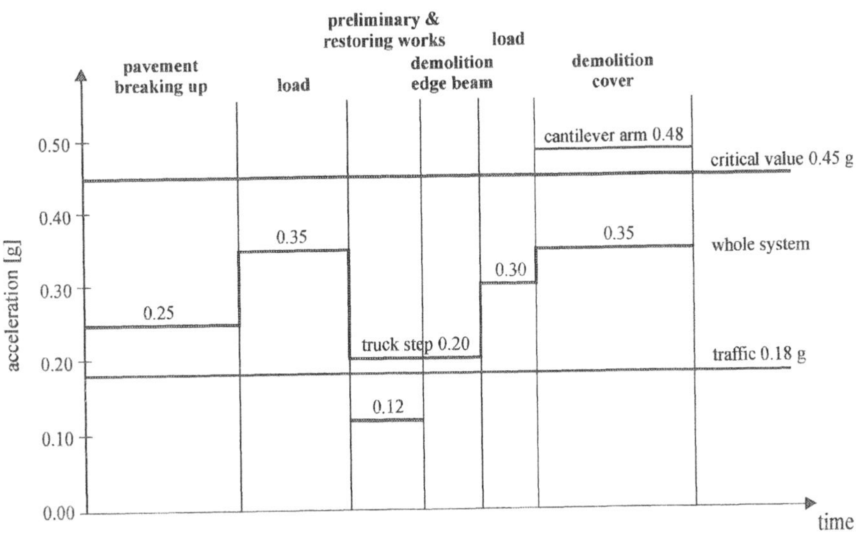

Figure 8.28 Loading of the structure over time

8.9 FLYOVER ST MARX, PERMANENT MONITORING

Location:	Vienna, Austria
Operator:	MA 29 (Bridge Maintenance by Magistrate of City Vienna)
Start of SHM:	November 1998
Structure category:	multiple span bridge: total length $L = 2700$ m
Spans:	54 column sets and 24 expansion joints between single bridge girders
Structural system:	pre-stressed and reinforced concrete box girder
Number of sensors installed:	4 × accelerometer, 1 × temperature sensor
Instrumentation design by:	VCE (Vienna Consulting Engineers), Austria

The St Marx Bridge (Figure 8.29) in Vienna, Austria, built from 1973 to 1978, is located between the Danube Canal and the Traffic Node Landstrasse. The bridge is counted among one of the most frequented partitions of the A23 South-East Highway. The total traffic volume averages about 240 000 motor vehicles per day, whereas an increase in the ratio of heavy loads is detected as well. This leads to an enormous loading of the bridge structure. As a consequence thereof the serviceability of the expansion joints and the bridge bearings is affected. Thus, in order to detect the passing heavy loads, which cause damage, a structural health-monitoring system (Tables 8.5 and 8.6 and Figure 8.30) in combination with a video control system have been installed in 1998. Furthermore, the remaining structural service life can be predicted as well by data analysis procedures listed in Table 8.7.

The substructures of consideration, namely TW4 respectively TW5, with a total length of 205.10 m represent a seven-span continuous beam with 29.30 m span length respectively. The cross-section, however, comprises three lines with a width of 3.25 m, whereas the total width is 12.88 m. The construction type is a pre-stressed concrete box girder with dimensions 1.96 × 4.50 m.

Figure 8.29 St Marx Bridge in Vienna

Table 8.5 Sensor details

Type of sensors	Number	Location
Accelerometers	4 channels at 2 sensors per superstructure	At the box girders of the spans 1 and 2 (at $0.6 \times L_{span}$ from the span beginning)
PT100	1 at substructure TW5	At the box girder of the first span (at the beginning)

Table 8.6 Measurement equipment and data management

Type of system	Data management
PC-based measurement system	Data pre-analysis (statistics, system identification) Main analysis, graphical presentation and documentation in office Data transfer via modem Long-term database

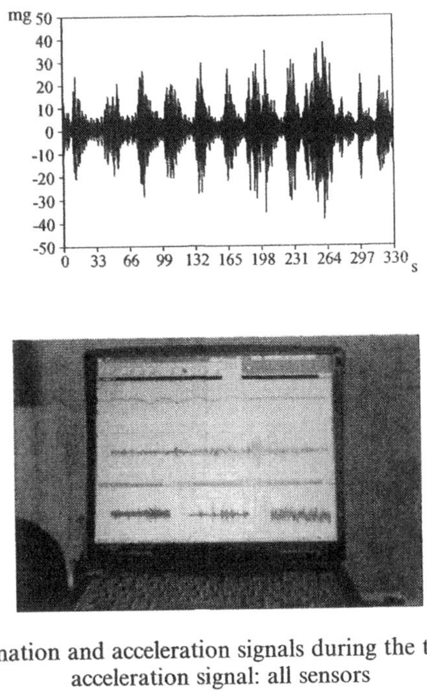

Figure 8.30 Deformation and acceleration signals during the test phase and vertical acceleration signal: all sensors

Table 8.7 Data analysis procedures

Type of analysis	Software	Additional features
Time domain and frequency domain SI, statistics, damage detection, FEM update, lifetime prediction	Self-made software, Octave 2.1.50, RSTAB 5.13.042, ANSYS 5.3	No expert system

On the basis of a permanent analysis of the dynamic structural behaviour possible issues to be considered are as follows:

- determination of passing heavy loads causing structural damage;
- verification update of the existing numerical load models;
- determination of the overall load configurations and vibration coefficients, whereas wind and temperature effects are considered optionally;
- consideration of long-term trends with respect to the life loads by means of statistics;
- monitoring of the structural loading capacity and serviceability by means of structural identification.

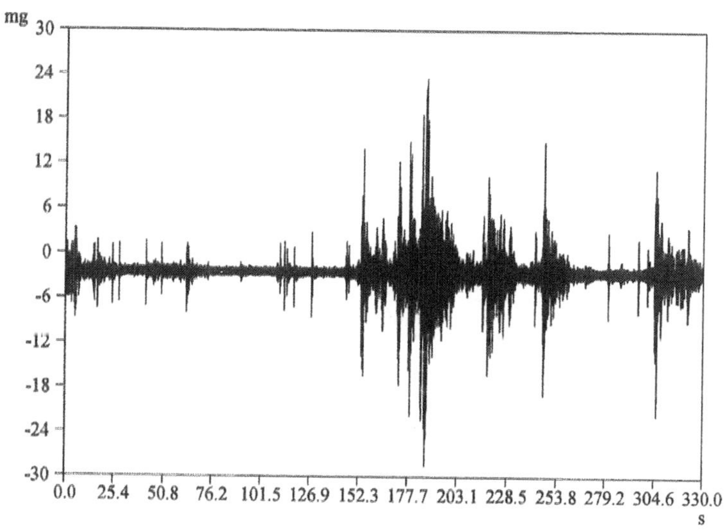

Figure 8.31 Vertical acceleration due to a vehicle crossing at time instant 150 s

Examples of outcomes:

Using the obtained accelerations time–history data system (Figure 8.31) identification is carried out by time domain as well as frequency domain methods. In general, the observed bridge structure is characterized by a distinct dynamic behaviour (Figure 8.32). Therefore a long-term SHM is very applicable. The implemented statistic analysis showed a relevant influence of the heavy loads. Environmental effects, e.g. wind-induced vibrations and temperature influence (Figure 8.33), are recognized as well. Additionally, in order to simulate the structure numerically and to detect damage the FE model update (Figure 8.34) is applied.

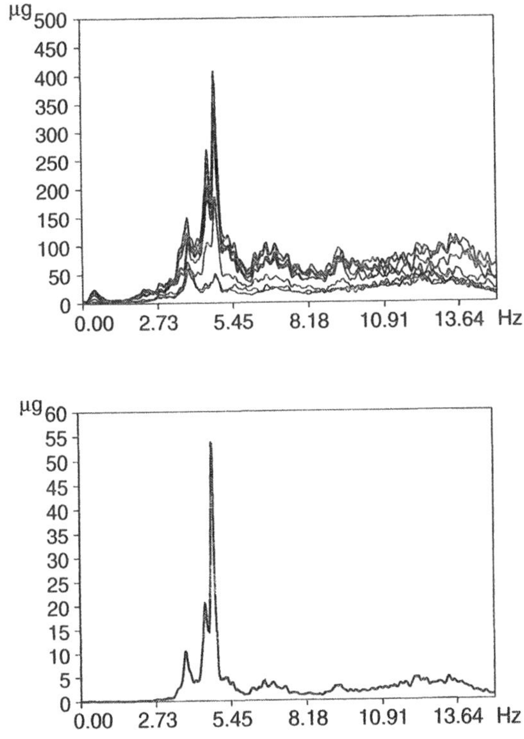

Figure 8.32 Spectral analysis of all sensors and ANPSD

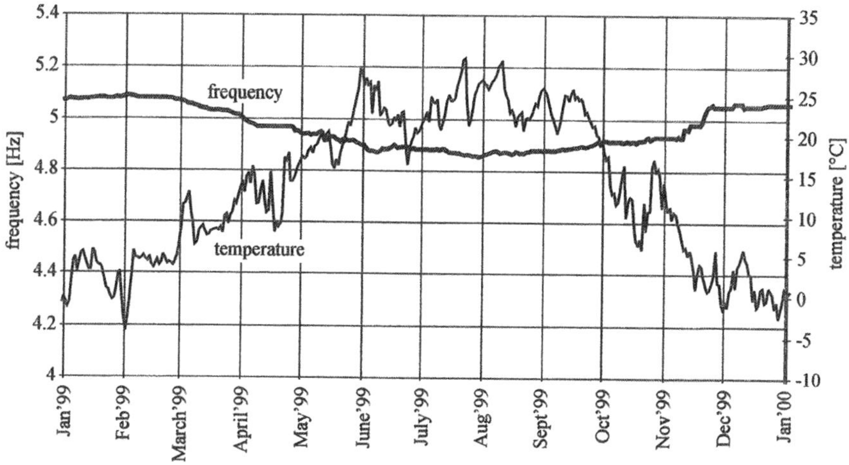

Figure 8.33 Temperature–frequency relationship over 1999

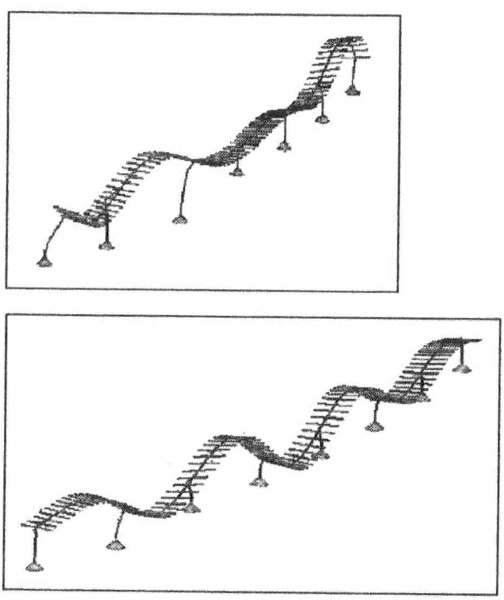

Figure 8.34 First vertical mode shape of TW4 and TW5

8.10 MUR BRIDGE IN ST MICHAEL, BRIDGE REHABILITATION

Location: St Michael, Styria, Austria
Client: ÖSAG–Austrian Motorway and Thruway Corporation
Checking period: 1998

In the course of carrying out a special assay of the load-bearing structure of the Mur Bridge (Figure 8.35), with the object to clear up bending in the main opening, the actual security of the pre-stressed concrete system was checked. By applying an additional external pre-stress the observance of admissible stress and sufficient structural safety could be verified.

The system consisting of two lanes is a five-span pre-stressed concrete structure with an overall length of 329 m. The maximum length in the central span amounts to 105 m and the construction width is about 17.3 m. The analysis was carried out to check the existing structural safety/load-bearing capacity whereby the external pre-stressing elements were submitted to a detailed analysis (Figures 8.36 to 8.39). It is therefore possible to assess the actual value of the pre-stressing force inside the cables in an easy and fast way.

Figure 8.35 Mur Bridge

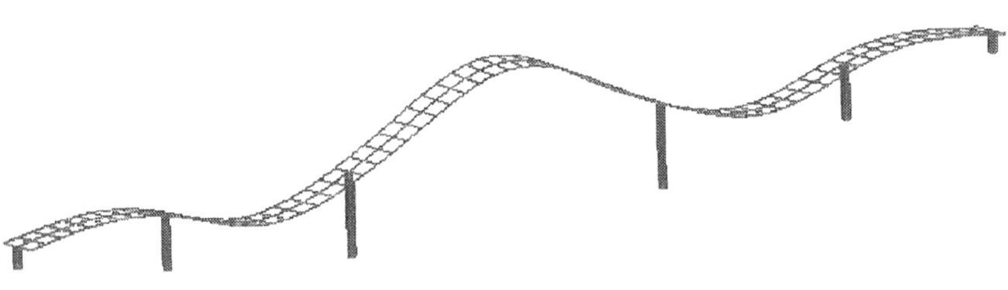

Figure 8.36 First vertical mode shape of the structure (calculation)

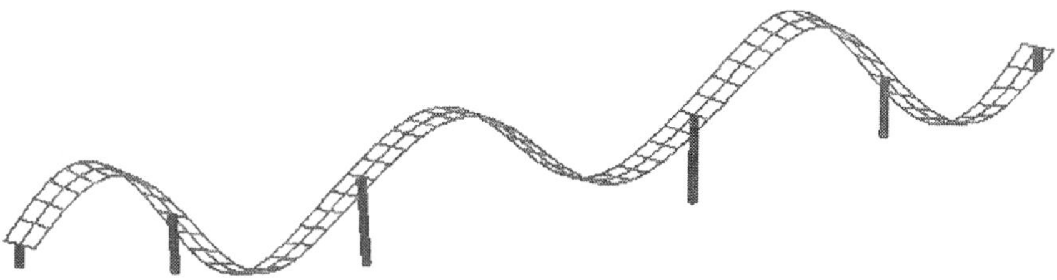

Figure 8.37 Second vertical mode shape of the structure (calculation)

Figure 8.38 Installation of sensors on CMM cables (marginal fields)

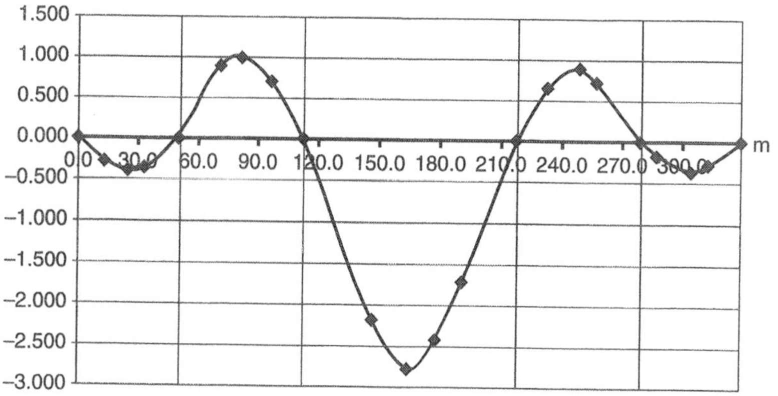

Figure 8.39 Vertical eigenform of the structure (measurement)

8.11 ROSEN BRIDGE IN TULLN, CONCRETE CABLE-STAYED BRIDGE (1995)

Location:	Tulln, Lower Austria, Austria
Client:	Government of Lower Austria
Construction period:	1992–1995
Checking period:	since 1998 every year

The Rosen Bridge (Figure 8.40), the new cable-stayed bridge across the Danube, was finished in 1995. The maximum span is 177 m and the bridge consists of a 70 m high A-shaped pylon. Tests have been performed already during construction and after completion of the bridge.

The system made it possible to keep track of all the changes induced by the construction stages and to show the redistribution of loads in the cables due to creep and shrinkage effects in the concrete bridge. Determination of the actual cable forces (steel) via frequency measurement (Figures 8.41 and 8.42) is a well-established method, which provides reliable results (Table 8.8). A minor difficulty is the correct interpretation of the effect of the grouting.

The final results of the dynamic measurements of the cables show that there is a very good correlation between theory and practice. The relation between eigenfrequencies of cables and cable forces by the well-known experience is

Figure 8.40 Rosen Bridge in Tulln

$$f_k = \frac{k}{2l}\left(\frac{N}{m}\right)^{\frac{1}{2}} \tag{8.1}$$

where
$k = 1, \ldots, n$
N = cable force
m = mass of the cable per metre length
l = length of the cable

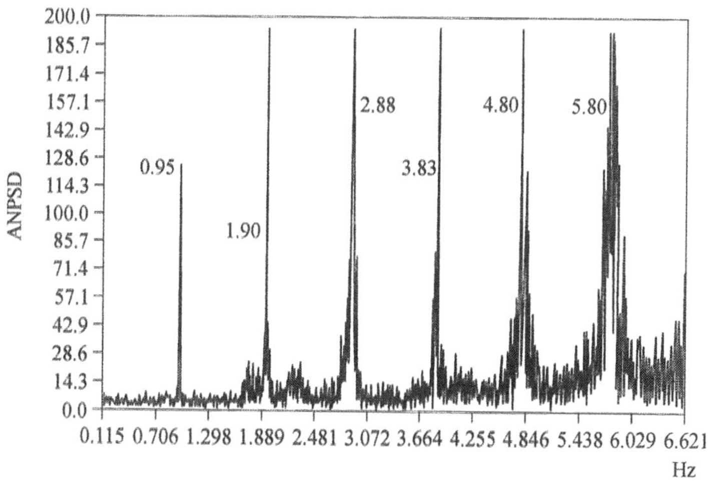

Figure 8.41 Measurements of natural frequencies under regular traffic

Figure 8.42 Installation of sensors on the cable

Table 8.8 Table of eigenfrequencies of stay cables (downstream side)

Cable number	Cable type	Eigenfrequency measured (Hz)	Cable length between anchorage (m)	Cable length between dampers (m)	g injection (kg/m)	g steel (kg/m)	g total (kg/m)	Cable force measured (kN)	Cable force determined by hydraulic jack (kN)
1	24–150	2.08	48.57	44.45	30.88	28.26	59.14	2022	1994
2	24–150	2.07	45.67	41.40	36.13	28.26	64.39	1892	1889
3	24–150	1.58	60.96	56.45	30.35	28.26	58.61	1865	1849
4	32–150	1.33	67.57	62.82	37.00	37.67	74.67	2085	1968
5	26–150	1.37	74.53	69.53	29.52	30.62	60.14	2183	2191
6	32–150	1.24	81.73	76.47	37.32	37.67	74.99	2697	2640
7	36–150	1.15	89.09	83.56	35.36	42.39	77.75	2872	2798
8	34–150	1.11	96.52	90.71	36.26	40.04	76.30	3094	3023
9	36–150	1.03	103.99	97.91	35.10	42.39	77.49	3152	3090
10	36–150	0.99	111.41	105.08	35.01	42.39	77.40	3351	3289
11	36–150	0.95	118.73	112.16	34.95	42.39	77.34	3512	3462
12	36–150	0.89	125.86	119.06	34.96	42.39	77.35	3474	3396
13	34–150	0.86	132.72	125.69	36.17	40.04	76.21	3562	3516
14	30–150	0.76	139.25	131.99	38.42	35.33	73.75	2968	2983
15	26–150	0.74	145.36	137.91	29.58	30.62	60.20	2508	2482

8.12 VOEST BRIDGE, STEEL CABLE-STAYED BRIDGE (1966)

Location: Linz, Upper Austria, Austria
Client: Government of Upper Austria
Checking period: 1999

The VOEST Bridge (Figure 8.43) across the Danube in Linz is a six-tracked road bridge with pavements on both sides. It serves as a flyover for the A7 Mühlkreis motorway. The basic system is a central beam cable-stayed bridge with one pylon constrained in the division of the girder and three parallel cables conducted over the pylon. The guying is asymmetrically arranged. The effective span lengths amount to $2 \times 60 + 72 + 215$ m and the total width of the structure is 34.9 m. The girder is a continuous steel girder with an orthotropic deck. The stiffening girder consists of four main girders with distances of 8.4 m and a height between 3.8 m and 5.5 m. The pylon has an overall dimension of 2.6×3.4 m in the height of the deck, which tapers lightly to the top until its total height of 65.0 m. Results of SHM are shown in Figures 8.44 to 8.51.

Figure 8.43 VOEST Bridge

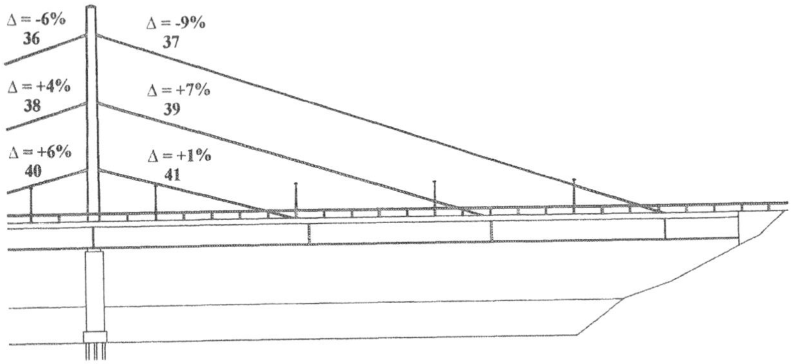

Figure 8.44 Losses in pre-stressing forces of all cables

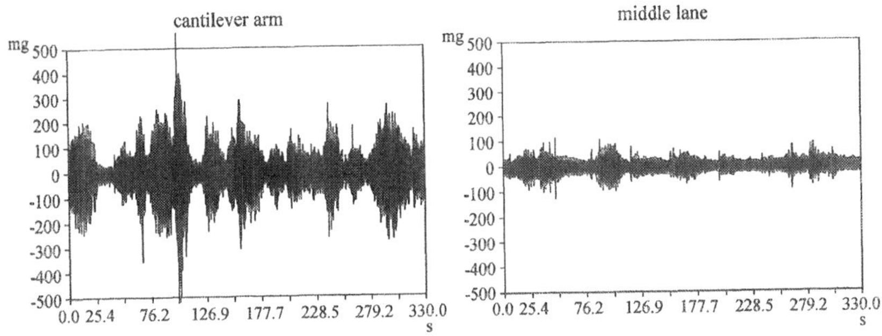

Figure 8.45 Comparison of acceleration level

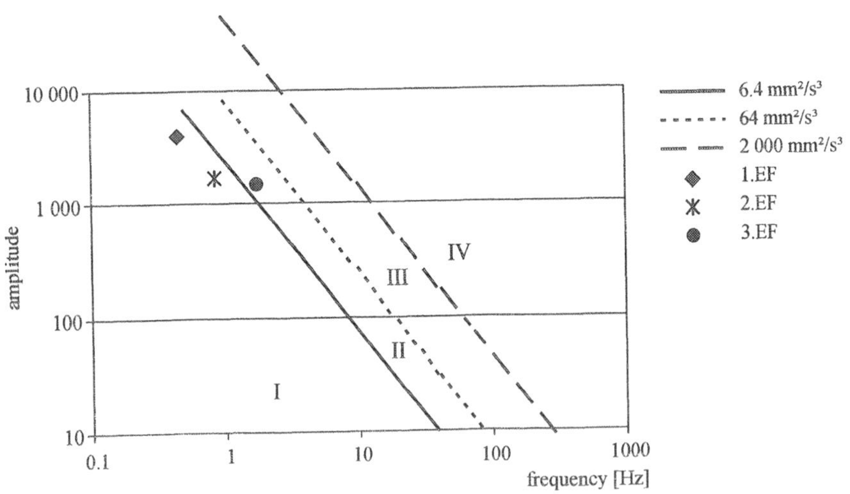

Figure 8.46 Respective vibration intensity: cantilever arm

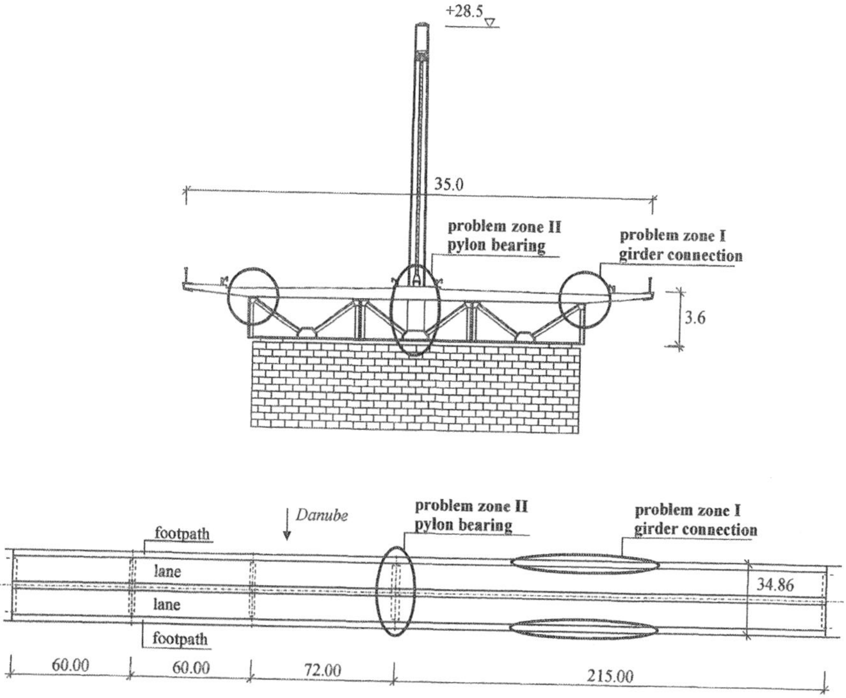

Figure 8.47 Problem zones of the structure (all dimensions in m)

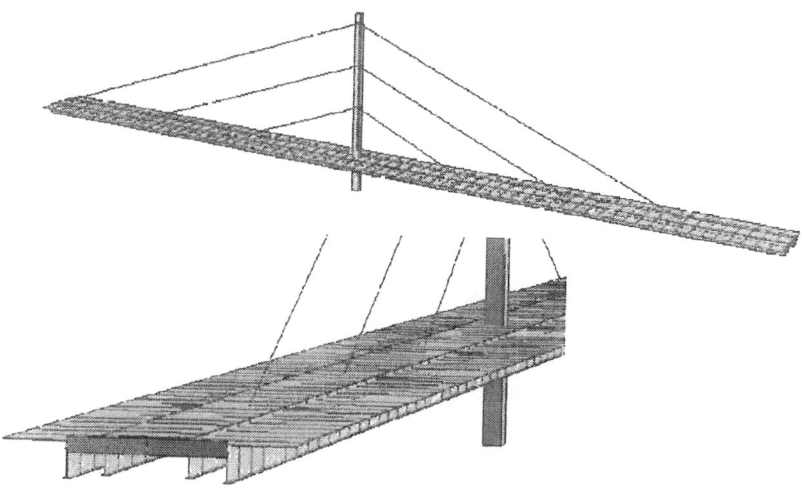

Figure 8.48 FE model of the structure

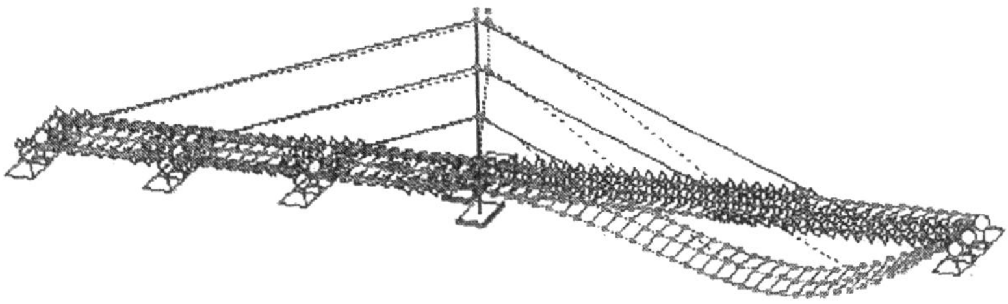

Figure 8.49 First vertical mode shape of the structure

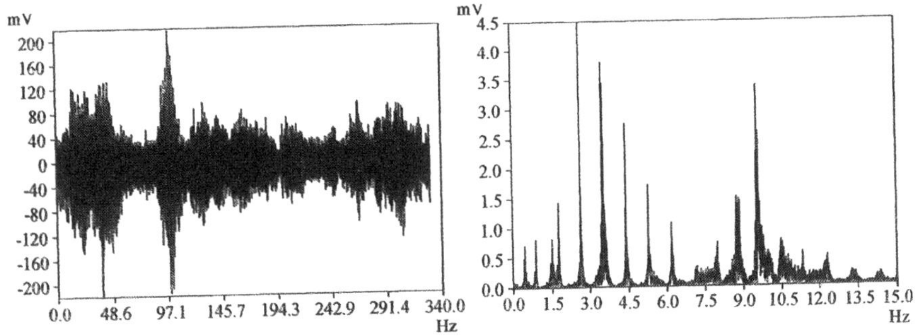

Figure 8.50 Acceleration signal and frequency spectrum of a cable

Figure 8.51 Installation of sensors on the cable

8.13 TAICHUNG BRIDGE, CABLE-STAYED BRIDGE

Location:	Taichung, Taiwan
Operator:	BPI Taiwan
Start of SHM:	November, 2003
Structure category:	cable-stayed bridge
Cables:	44
Spans:	2 spans: 89.5 / 89.5 m
Height of the pylon:	80 m
Structural system:	steel girder with orthotropic deck
Number of sensors installed:	15
Instrumentation design by:	VCE (Vienna Consulting Engineers), Austria

The Taichung Bridge (Figure 8.52), opened in 2003, is a cable-stayed bridge for urban traffic located in Taichung in the middle of Taiwan. Owing to the requirement to assess the cable forces, the global state of the structure and the dynamic behaviour of the pylon base, a permanent monitoring system was installed in 2003.

The Taichung Bridge is a cable-stayed bridge with 44 cables and a total length of 189 m, which comprises four lanes and two small lanes for pedestrians and bicycles. The superstructure is represented by steel girders and an orthotropic deck.

Figure 8.52 Taichung Bridge, Taiwan

The permanent monitoring system (Tables 8.9, 8.10 and 8.11) gives an overview of the global behaviour of the bridge structure and supplies the actual cable forces. The system consists of following parts, which are monitored:

- dynamic determination of the cable forces of eight selected cables;
- measuring of temperature, wind speed and wind direction;
- dynamic measurement of the main girders and the pylon top;
- three-dimensional measurement of the pylon base.

Examples of outcomes:

The permanent monitoring system at Taichung Bridge measures vibration, temperature and wind. The self-made software supplies the cable forces of eight selected cables in the way that the client can easily check the status of the cable forces in the form of a signal light (Figures 8.54 and 8.55).

Table 8.9 Sensor details

Type of sensors	Number	Location
Acceleration transducers	8	At 1 cable each
Velocity transducers	3	At the main girders
Three-dimensional acceleration transducer	1	At pylon base
Wind sensors (Figure 8.53)	1	5 m above the road surface
Temperature sensors	2	Inside and outside the box girder

Table 8.10 Measurement equipment and data management

Type of system	Data management
PC- and standalone-based measuring system	Storage in a long-term database on site Analysis (statistics, frequency analysis, etc.) and graphical presentation and documentation in office Controlling of the successful operation of the measuring system via the modem

Table 8.11 Data analysis procedures

Type of analysis	Software	Additional features
Ambient analysis, calculation of cable forces and lifetime calculations	Self-made software	No expert system

Figure 8.53 Wind sensor at Taichung Bridge, Taiwan

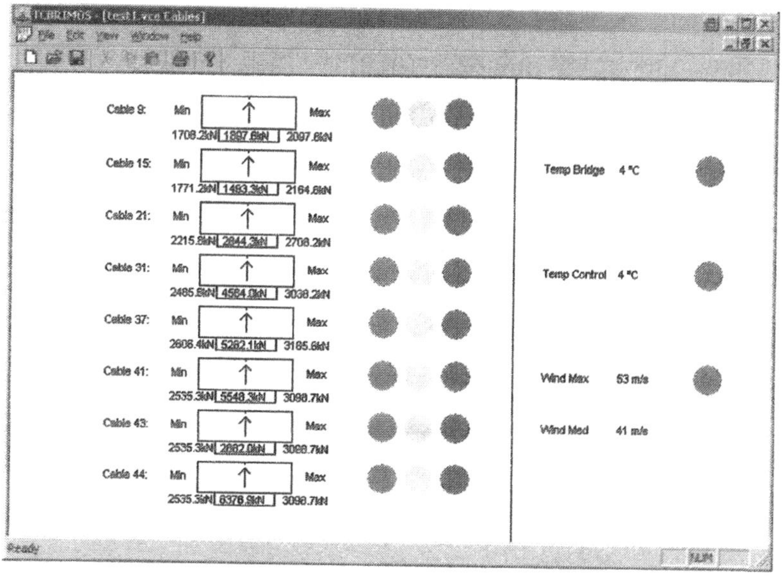

Figure 8.54 Theoretical output of the monitoring system

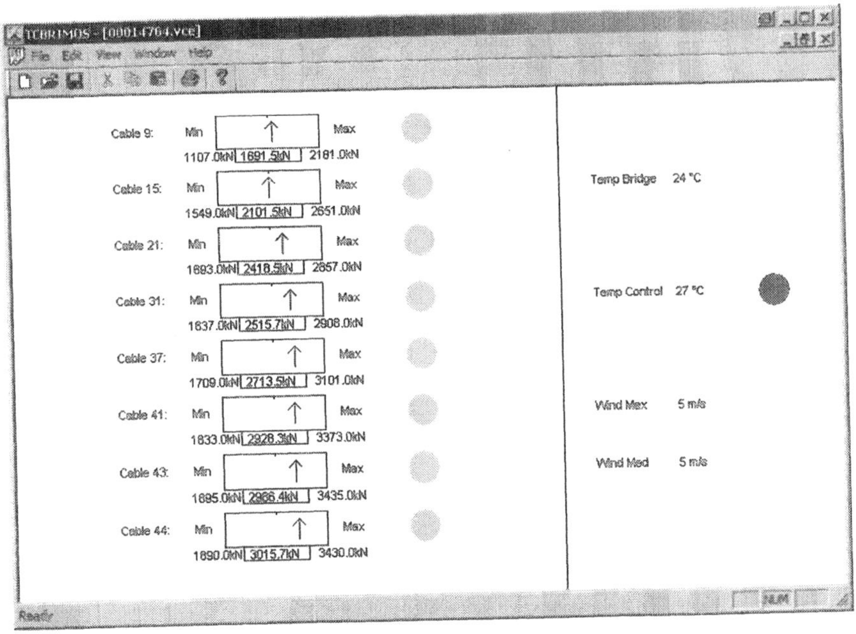

Figure 8.55 Real output of the permanent monitoring system (the green light shows immediately that all cable forces are all right)

Benefits of using permanent measuring system in the project:

The ability to merge high-precision sensor data of accelerations and velocities in dependence of separately registered wind and temperature data provides the possibility to realize lifetime considerations, which are of the highest importance for bridge operators.

Appendix

NOMENCLATURE

a	acceleration, coefficient
a^e	dimensionless correction factor
$\boldsymbol{A}, \boldsymbol{B}, \boldsymbol{C}, \boldsymbol{D}$	discrete-time state-space model
b	coefficient
c	damping, stiffness
$\boldsymbol{C}$	constant value, damping matrix
d	diameter
d_{dyn}	dynamic deformation
d_{stat}	static deformation
D	damping according to Lehr
D_i	partial damages
e	element
e_k	white noise term at time instant k
E	expected value operator, Young's modulus
$\boldsymbol{E}$	covariance matrix
EI	bending stiffness
f	frequency (Hz)
f_{i}	inertial force
F	force
$\boldsymbol{G}$	'next state-output' covariance matrix, shear modulus
$h(t-\tau)$	unit-impulse response function
H	horizontal force

Ambient Vibration Monitoring H. Wenzel and D. Pichler

$\boldsymbol{H}^{\mathrm{ref}}$	output data block Hankel matrix
$H(f)$	transfer function
$H(\mathrm{j}\bar{\omega})$	complex frequency response function
i	imaginary unit
i	indices
I	moment of inertia
$\boldsymbol{I}(\mathrm{j}\bar{\omega})$	impedance matrix
j	complex number, imaginary unit, indices
$\boldsymbol{J}$	sensitivity matrix
$\boldsymbol{J}_\theta$	Jacobian matrix
k	galloping criterion parameter, indices, number
k_1, k_2	enlargement factors
K	Kalman gain
$\boldsymbol{K}$	stiffness matrix
K_{B}	vibration coefficient regarding DIN 4150 ÖNORM S9010
$\boldsymbol{K}^{\mathrm{e}}$	updated element stiffness matrix
$\boldsymbol{K}_0^{\mathrm{e}}$	initial element stiffness matrix
$\boldsymbol{K}^{\mathrm{u}}$	unchanged element stiffness matrix
$K1$	conventional pre-stressed mono-block concrete sleepers
l	cable length, number of outputs
$\boldsymbol{L}$	selection matrix that selects the references from the outputs
L_{max}	maximum span of the structure
m	indices, mass per unit, slope, structural mass
$\bar{m}$	mass per unit
M	number
$\boldsymbol{M}$	mass matrix
M_{dr}	basic design moment
n	indices
n_e	number of updated elements
n_i	number of fatigue relevant loading cycles, number of occurring loading cycles
N	allowable number of cycles, cable force, number
$\mathcal{N}(\boldsymbol{x}, \boldsymbol{\theta})$	global damage function
$\boldsymbol{O}_i$	observability matrix of order i
p	indices, load
$\boldsymbol{p}$	multiplication factors
p_{eff}	effective load
$\boldsymbol{P}$	forward state covariance matrix
P_{d}	design value of the load on the sleeper
$\mathcal{P}_i^{\mathrm{ref}}$	projection of the row space of the future outputs into the row space of the past references
q	indices

$q_k(z)$	quadratic model function
$\boldsymbol{Q}, \boldsymbol{R}, \boldsymbol{S}$	process and measurement noise covariance matrices
Q_0	static load
r	damping, indices, number of references
$\boldsymbol{r}$	residual vector
$\boldsymbol{r}_f$	frequency residual vector
$\boldsymbol{r}_s$	mode shape residual vector
$\boldsymbol{R}$	covariance matrix
$\boldsymbol{R}_e$	covariance matrix of the innovations e_k
$\boldsymbol{R}(t)$	excitation vector
S	station of BRIMOS® recorder measured from a column of the middle span
Sc	Scruton number
t	time
$\boldsymbol{t}$	shape parameters
T	time period
$\boldsymbol{T}$	kinetic energy
u_k	input at time instant k
u_{measured}	measured value
u_{real}	measurable value
$\boldsymbol{U}(t)$	displacement vector
U_i	potential energy
v	displacement, velocity
v_k	measurement noise due to sensor inaccuracy
$\boldsymbol{V}$	vertical force
V_{crit}	critical wind speed
w	displacement
w_k	process noise due to disturbances and modelling inaccuracies
$\boldsymbol{W}$	weighting matrix
$W(\mathrm{j}\bar{\omega})$	displacement in the frequency domain
$\boldsymbol{x}$	geometrical coordinates
x_i	discrete measuring value
x_k	state at time instant k
x_{Trigger}	trigger value
X	variable
$\hat{X}_i$	Kalman filter state sequence
y_k	output at time instant k
$\boldsymbol{Y}_{\mathrm{p}}^{\mathrm{ref}}, \boldsymbol{Y}_{0\mid i-1}^{\mathrm{ref}}$	Hankel matrix of past reference outputs
$\boldsymbol{Y}_{\mathrm{f}}, \boldsymbol{Y}_{i\mid 2i-1}$	Hankel matrix of all future outputs
$Y(t)$	modal coordinate
$z_j(\theta)$	analytical modal quantity
$\tilde{z}$	measured quantity
$Z_n(t)$	nth modal coordinate

β	ratio
γ_{i}	dynamic increment for uncertainties in the contact sleepers – ballast bed
γ_{d}	load distribution factor
γ_{p}	dynamic increment for damping effects
γ_{r}	partial safety factor (covers uncertainties in the contact sleepers – ballast bed)
γ_{v}	regular dynamic increment (covers speed effects)
δ	real number
$\delta_{\mathrm{absolute}}$	absolute measurement error
δ_{pq}	Kronecker delta
δ_{real}	relative measurement error
$\delta(t)$	Dirac delta function
$\Delta\sigma$	certain stress range level
$\Delta\sigma_D$	constant amplitude fatigue limit
$\Delta\sigma_L$	cut-off limit
ζ	damping coefficient
θ	optimization variable
λ	characteristic root, eigenvalue
ξ	damping ratio, related bending stiffness
ρ	density
σ	normal stress
τ	time instant of applied impulse
ϕ	dynamic amplification factor
$\phi_n(x)$	*n*th mode shape function
ψ	mode shape
ω	natural circular frequency, real number
ω_{D}	damped natural circular frequency
ω_e	natural circular frequency
ω_i	natural circular eigenfrequency
$\bar{\omega}$	loading circular frequency
$[\cdot]$	matrix, vector
$(\cdot)^{\mathrm{T}}$	transpose
$\cdot^{\dagger}$	Moore–Penrose pseudo-inverse
$\cdot(\mathrm{j}\bar{\omega})$	frequency domain variable
$\cdot(s)$	space domain variable
$\cdot(t)$	time domain variable
$\cdot(x,t)$	variable over place and time
$\cdot_{\mathrm{h}}$	homogeneous solution
$\cdot_{\max}$	maximum value
$\cdot_{\min}$	minimum value
$\cdot_{\mathrm{p}}$	particular solution
$\cdot_{,\mathrm{x}}$	first derivative with respect to the place

$\bullet_{,xx}$	second derivative with respect to the place
$\dot{\bullet}$	first derivative with respect to the time $\hat{=}$ velocity
$\ddot{\bullet}$	second derivative with respect to the time $\hat{=}$ acceleration
$\ddot{\bullet}_g$	ground acceleration
$\tilde{\bullet}$	experimental value
det $(\cdot)$	determinant
exp $(\cdot)$	exponential
$f(\cdot)$	function
$F(\cdot)$	function
ln $(\cdot)$	natural logarithm
$\mathcal{L}\{\cdot\}$	Laplace transform

ADTV	Average Daily Traffic Volume
ANPSD	Averaged Normalized Power Spectral Density
APSD	Averaged Power Spectral Density
AVD	Ambient Vibration Derivatives
AVM	Ambient Vibration Method
BRIMOS®	Bridge Monitoring System
CQC	Complete Quadratic Combination
DFT	Discrete Fourier Transformation
EMPA	Eidgenössische Materialprüfungs -und Forschungsanstalt
FAMOS	Fast analysis and Monitoring Of Signals
FEM	Finite Element Method
FFF	Forschungs förderungs fond
FFT	Fast Fourier Transformation
FR	Fatigue Relevance
LSCE	Least Squares Complex Exponential
MAC	Modal Assurance Criteria
MDOF	Multi-Degree-Of-Freedom
MSS	Mass–Spring System
RDT	Random Decrement Technique
RMS	Root Mean Square
RS95	New frame sleepers
SDOF	Single-Degree-Of-Freedom
SHM	Structural Health Monitoring
SI	System Identification
SLD	Sylodyn®
SLS	Sylomer®
SRSS	Square Root of the Sum of the Squares
SSI	Stochastic Subspace Identification
SSI-DATA	DATA-driven Stochastic Subspace Identification
SVD	Singular Value Decomposition
VCE	Vienna Consulting Engineers
VCM	Vibration Characteristic Method

Index

Note: Page numbers appearing in bold refer to figures and page numbers appearing in italic refer to tables.

Ambient Vibration Monitoring H. Wenzel and D. Pichler

Printed and bound by CPI Group (UK) Ltd, Croydon, CR0 4YY
16/09/2026
14963379-0008